BELGIAN MILITARY AVIATION 1945-1977

compiled by Paul A. Jackson

Midland Counties Publications

PUBLISHED BY
Midland Counties Publications
17 Woodstock Close
Burbage, Hinckley, Leicestershire
LE10 2EG, England

ISBN 0 904597 06 7 (Paperback)
0 904597 05 9 (Hardback)

PRINTED BY
W. T. Carrick (Printers) Limited
Westminster Trading Estate
Measham, Leicestershire
England

Contents

FRONT COVER

Dassault Mirage 5BA BA-37, photographed at Kleine Brogel in May 1976, serves with 1 Smaldeel.
Photograph by S.G. Richards.
Cover design and artwork preparation by Warwickshire Illustrations.

Introduction pour les Lecteurs Francophones

Ce livre est le troisième de la série que publie Midland Counties Publications et qui donne des informations détaillées sur les forces aériennes Européennes.

La recherche de documents concernant l'histoire des avions, leurs insignes et les escadrons où ils ont opéré, est populaire en Grende-Bretange depuis de nombreuses années et elle s'est étendue a d'autres parties de l'Europe, particulièrement aux Pays-Bas et récemment en France. Des ouvrages de ce genre sur les avions civils ont été publiés depuis 1951 mais l'augmentation de liaisons efficaces entre les 'spotters' militaires, grâce à des revues de passionnés telles 'British Aviation Review', 'Flash', 'South-East Air Review' et 'Trait d'Union', a rendu disponible des informations nécessaires pour ce livre.

La majorité des lecteurs étant de langue anglaise, pour simplifier, nous avons donné tous les noms en une seule langue. Nous espérons que les Belges de langue française nous excuserons d'utiliser le Flamand, parce que la majorité des noms des escadrons et bases sont plus faciles à comprendre et à parler pour un Anglais (par exemple: 'school' en flamand reste 'school' en anglais mais devient 'école' en français).

Nous espérons que vous trouverez ce livre très utile pour votre passe-temps préféré et si vous possédez n'importe quelle information, l'auteur sera heureux de la recevoir pour en faire bénéficier d'autres passionnés.

Inleiding voor Vlaams-Sprekende Lezers

Dit boek is het derde in de serie van gedetaillieerde informatie over de Europese luchtmachten gepubliceerd door Midland Counties Publications.

Het vastleggen en bijhouden van informatie over vliegtuigen, hun merktekenen en de smaldelen waarbij zij hebben gediend is al sinds vele jaren erg geliefd in England en is ook overgewaaid naar andere delen van Europa, in het bijzonder naar Nederland en onlangs ook naar Frankrijk.

Over burgervliegtuigen worden al sinds 1951 boeken gepubliceerd, de militaire informatie benodigd voor dit boek kon worden bijeengebracht door efficiënte samenwerking tussen de militaire 'spotters', middels magazines zoals daar zijn 'British Aviation Review', 'Flash', 'South-East Air Review' en 'le Trait d'Union'.

We hopen dat U nuttig gebruik kunt maken van dit boek en mocht U nieuwe informatie ontdekken, de auteur zou er erg blij mee zijn, zodat het gebruikt kan worden ten voordele van andere enthousiasten.

Introduction & Acknowledgements

In this, the third of a series of books reviewing Europe's military air arms, the subject is Belgium and its neighbouring Grand Duchy of Luxembourg. The two states are normally associated with the Netherlands under the convenient umbrella title of 'Benelux', but in order that they may receive full coverage, the three will be dealt with in two publications.

It is the intention of this book to present as accurately as possible, a picture of military aviation in Belgium in 1977. This has been coupled with historical data giving details of aircraft flown in military markings since 1940, including wartime RAF squadrons.

As with the previous books in this series, technical specifications and information are deliberately kept to a minimum, as these are readily available to those requiring them from such publications as 'Janes' and the 'Observers Books'.

Likewise we make no pretence that this book contains the full picture; this can only emerge on the release of official records in many years time.

The chequered political history of Belgium is responsible for the present existence of two official languages. The nine provinces are divided between the Flemish speaking North and French speaking South, whilst some German is also spoken in the East. Luxembourg speaks both French and German. As the majority of readers are expected to be English-speaking, this work has standardised on Flemish proper nouns throughout, for simplicity. No offence is intended towards Belgians speaking other languages, and a glossary is on page 8.

Contents of the following pages are mainly from the author's personal records, but have been greatly enriched by contributions from several individuals. A wealth of information has been provided by Valentijn Kenens to whom I extend grateful thanks. Others who have been of assistance over the years have been: F. Binnemans, the editors and contributors of British Aviation Review, the editors and contributors of Flash, Ray Sturtivant, Terry Sykes, the editors and contributors of South East Air Review and not least, Johan van der Wei.

Special mention must be made of Dave Sargent who has removed a heavy burden from the author's shoulders by undertaking the drawing of all unit badges, sometimes from freehand originals or grainy photographs. Aircraft illustrations have been carefully selected from many hundreds submitted, and it is a source of regret that it has not been possible to include them all. Photographs, like written information, are often exchanged between enthusiasts and advance apologies are offered if the odd one has been incorrectly credited.

I shall be pleased to receive amendments to the information contained in the following pages at the address below. Readers who wish to keep track of new developments or who have access to hitherto unpublished information which will be of interest to fellow enthusiasts should refer to the appendix 'Keeping up to date' for the addresses of amateur magazines which regularly contain European information.

Paul A. Jackson September 1977
50 Wilby Park, Wilby
Wellingborough, Northants
NN8 2UL

Abbreviations

The abbreviations used in this book consist mainly of those familiar to most readers, viz:

c/n = constructor's number; w/o = written off;
wfu = written off.

However there are occasions where unit designations have been abbreviated, for instance Sm = Smaldeel, Smn = Smaldelen, W = Wing, Sq = Squadron. Please also see the introductions to the individual sections, where applicable.

Warning!

The hobby of aircraft spotting can be misunderstood in countries where it is not as widely accepted as in the UK and this has lead to a number of unfortunate incidents over the past few years.

We therefore remind readers of the advice contained in the official booklet *Essential Information for British Passport Holders:* 'Hobbies like *aircraft*, train or ship spotting, and even bird watching, are liable to misinterpretation and may lead to your being arrested for spying in some countries abroad. If in doubt you should enquire from the local authorities or a British consular officer'.

Part One

A History

Historical Background

To understand Belgium's present-day position in Europe - not least its dual language situation, it is necessary to go back in time to the Treaty of Verdun in 843 under which the Carolingian Empire was disposed. Having previously been outposts of the Roman and later Frankish Empires, the European lowlands gained autonomy as Lotharingia (with the exception of Flanders which remained French, but with a large degree of self government). Lotharingia attached itself to the German states after 925, but internal disputes ultimately resulted in the formation of the Duchies of Brabant, Limburg and Luxembourg, and the Counties of Hainaut, Holland, Namur and Zeeland, plus the Principalities of Liege and Utrecht.

Subsequent re-integration of these areas built up the Netherlands (roughly equivalent to Benelux) to which were added Burgundy, Franche-Comte, Flanders and Artois (the latter all in modern northern France). By the Union of Utrecht recognised in 1648, the seven northern (Protestant) provinces achieved self-determination whilst the rest remained under Spanish (Catholic) rule.

During the eighteenth century, Belgium was a part of the Austrian Empire known as the Austrian Netherlands. A United States of Belgium was formed in 1790, but this joined with France two years later and was formally annexed in 1795.

Following the defeat of Napolean the country was occupied by British, Prussian and other Allied forces in 1813 and amalgamated by the Congress of Vienna with Holland as the Kingdom of the Netherlands in 1814. At this time, Luxembourg allied itself to the German Confederation, later becoming an independent Grand Duchy in 1867.

After the revolt of 1830, the Southern half of the Netherlands Kingdom broke free and established what is now Belgium, under Leopold I. An agreement of the European powers, the Treaty of XXIV Articles, was signed on 15 November 1831, guaranteeing Belgium sovereignty and neutrality. The title 'Netherlands' was retained by the remainder of the Kingdom and with the exception of Luxembourg's economic union with Belgium in 1921, the position has remained static.

The Flemish Language

The chequered political history of Belgium is responsible for the present existence of two official languages, Flemish and French. The two ethnic groups have maintained their separate identities, and a broad generalisation divides them as Flemings (Protestants from the North) and Walloons (Catholics from the South, mostly French-speaking).

The Flemish and Dutch languages share a common vocabulary, but have differences in pronunciation, as do regional accents in Britain. In view of the closeness of the two communities, both groups have at least a working knowledge of the other's language, but the British visitor using French (or Dutch) is advised to state his nationality when addressing a Belgian in order not to appear discourteous if he is using the 'wrong' language.

In order to impart a degree of uniformity with the second section of this book, Flemish proper nouns will be used throughout, but accompanied by a table of their French and English equivalents.

No such simplistic answer to the dual language problem can be applied by the Luchtmacht, and thus, bases north of Brussels are Flemish-speaking whilst those to the south are French. Training units and the transport wing are dual. Below is a list of current bases with their different spellings and the languages spoken.

Flemish	French	Language
Bevekom	Beauvechain	dual
Bierset	Bierset	French
Brustem	Brustem	dual
Brasschaat	Brasschaat	dual
Florennes	Florennes	French
Goetsenhoven	Gossoncourt	dual
Koksijde	Coxyde	dual
Kleine Brogel	Petit Brogel	Flemish
Melsbroek	Melsbroek	dual

The following terms appear regularly in this book and a comparative translation is given in three languages. In a few instances the 'RAF equivalent' is given in place of the literal translation.

Flemish	French	English
Alle Weder	Tout temps	All-weather
Belgische Luchtmacht	Force Aérienne Belge	Belgian Air Force
Dag jacht	Chasse de Jour	Day fighter
Elementaire Vliegschool	Ecole de Pilotage Elémentaire	Elementary Flying School
Groepering Opleiding en Training	Groupement de l'Instruction et de l'Entrainment	Instructional and Training Group
Groepering Tactische Luchtmacht	Commandement de la Force Aérienne Tactique	H.Q. Tactical Air Force
Jager	Chasse	Pursuit (fighter)
Jagers–Bommenwerpers	Chasseurs–Bombardiers	Fighter–bomber
Licht Vliegwezen	Observation Légère	Army observation
Luchtmacht	Force Aérienne	Air Force
Nacht	Nuit	Night
Smaldeel	Escadrille*	Squadron
Telegeleide Tuigen Grond – Lucht	Engin Téléguides Sol – Air	Ground – Air Guided Missile
Tactische	Tactique	Tactical
Vortgezette Vliegopleidings School	Ecole de Pilotage Avancé	Advanced Flying School
Verkennings	Reconnaissance–Photographique	Photo–reconnaissance
Vervoer en Verbindungs	Transport et Communications	Transport and Communications
Vlieg	Vol	Flight
Vliegen Zonder Zicht	Vol sans Visibilité	Instrument Flying
Wing†	Wing†	Wing
Zeemacht	Force Navale	Navy

Note that a third language, German, is spoken in the far east of Belgium, particularly the districts of Eupen and Malmedy which were awarded to Belgium in 1919.

*Not 'Escadron' which is the Armée de l'Air equivalent of squadron. In French service the Escadrille is equivalent in aircraft strength to a 'Flight' or half squadron.

†Adopted from RAF nomenclature 1939–45 and retained. Pre-war equivalent was 'Groep' or 'Groupement'.

The Belgian Air Force - A Short History

In common with most other European countries, the Belgian use of the air for military purposes began in the latter half of the nineteenth century. Although the balloon had been known for a hundred years, armies were slow to realise its potential. Military commanders of the Roman, and possibly earlier eras, had preferred to camp their armies on higher ground than their opponents, appreciating that this gave them the initiative not only in deployment, but in observing the position and preparation of the enemy.

The balloon was an extension of this principle, complementing, if not replacing the traditional role of the scouting cavalry. In contrast to its neighbours (France, Germany and Britain), Belgium had no territorial ambitions, and consequently organised its forces only on a defensive basis. Thus, when in 1887 General Brialmont established a balloon school and three operational sections of a Balloon Corps, they were equipped with Drachen balloons. The Drachen was a man–carrying German invention resembling the barrage balloon of the Second World War, designed only for tethered operations where its fins would keep it pointing into the wind and prevent it from rotating as would a sphere if not travelling freely through the air.

Heavier-than-air flying in the Belgian armed forces stems from a committee set up on 31 October 1910 to organise the establishment of a 'Compagnie van Werklui en Luchtschippers' (Aircraft crew and support corps). By 5 May 1911 the CWL had a strength of five pilots and three ground crew stationed at Brasschaat, and on that day its first aircraft, a Farman, was delivered . . . . and written-off in a crash. A replacement arrived on the 24th of the same month, and the Belgian air force was in business.

Recognition of the value of aviation was given on 16 April 1913 when a Royal Decree established the 'Compagnie van Vliegers' as a separate section of the Army known as the 'Militair Vliegwezen'. Four squadrons were formed, but when the First World War began the MV had only twenty five aircraft on charge – mostly Farmans with a few Deperdussin and Bleriot XI types, and 45 pilots.

Belgium was not directly involved in the political events which led up to the declaration of war, having earlier proclaimed its neutrality (as did Holland), but German designs on France envisaged an advance on Paris from the north, and Belgium was directly in the way. The Kaiser's army entered Belgium, and Britain, bound by treaty to protect the neutrality of the country, declared war on Germany on 4 August. Both sides anticipated a swift conclusion, but became embroiled instead in the stagnant trench warfare which characterised this conflict. By the time the trenches had been dug, only a small portion of Belgium in the south-west was free of German domination.

Early in 1915, the MV formed its sixth squadron (average complement five aircraft each), with three squadrons based at Koksijde and three at Houtem. German shelling of Koksijde forced a withdrawal to Moeren where the Farmans were replaced by more up-to-date Nieuports. Spads were received in 1917 together with other British and French built types, as the firm of Bollenens which had previously licence-built aircraft for the MV was in the German sector. Belgian crews additionally flew Short 827 seaplanes from Lake Tanganyika, where German colonies in Africa required the presence of containing forces.

When, on 11 November 1918 hostilities ceased, the MV had appoximately 120 aircraft in its eleven operational squadrons, mostly of French origin such as the Breguet 14, Farman F40, Hanriot HD1 and Spad S11. Little was done during the next few years to maintain the air force, apart from taking onto charge numbers of captured German aircraft.

By 1923, a thorough review was urgently required, and this resulted in the formation of several new squadrons, (bringing the total up to twenty-five) and the purchase of re-furbished wartime aircraft from France and Britain. These were augmented by licence production from the SABCA (Société Anonyme Belge de Constructions Aéronautiques) factory at Evère, and types involved were the Avro 504, Caudron G3, de Havilland DH4, DH9, Morane AS35, Nieuport 29C1 and Spad XIII.

In 1925, a further re-organisation transformed the Groeperingen into which the 25 squadrons had been divided, into regiments, No.1 – for reconnaissance and army co-operation, No.2 – fighter and No.3 – bomber. Nearly new though the aircraft purchased in 1923 were, the flesh was willing but, the spirit weak, and equipment of a more modern design was called for. In order to provide a little new blood, SABCA built the Avia BH-21 and Breguet XIX but little else was achieved by way of modernisation until 1931, when the government were made aware of the obsolete condition of the majority of the air force and authorised purchase of the Fairey Fox and Firefly.

The graceful Fox and Firefly, many of which were built by Avions Fairey, were augmented by a squadron of Gloster Gladiators in 1937, but it took the traumatic events of the next year to bring home to the government that not only was another war inevitable, but a declaration of neutrality would be as worthless as the piece of paper which Neville Chamberlain brought back from Munich.

Since the 1931 re-equipment programme, the operational biplane had been eclipsed by the monoplane in Britain, France and Germany and it was thus to the former two nations that Belgium looked for new designs such as the Hurricane, Oxford, Battle, and Breguet 694. Additionally, two squadrons of Brewster 339 were ordered from the USA, with Italy providing the Caproni 312, and Fiat CR42 and Holland the Koolhoven FK56.

When Britain and France declared war on Germany on 3 September 1939, Belgium was not immediately involved in the hostilities, and continued its re-armament programme. Despite receiving captured German documents detailing an attack on Belgium similar to that effected in 1914, the Belgians refused to allow any British or French forces to enter the country and bolster their defences.

The long expected blow came in the early hours of 10 May 1940 when Germany attacked Belgium, France and Holland in a concerted thrust to the North Sea coast. The first attack caught many aircraft on the ground, but the remainder fought valiantly on in defence of their homeland. On 11 May an Anglo-Belgian force of Battles attacked the bridges over the Albert Kanaal in an attempt to halt the flow of German troops. Severe losses were inflicted on the Battles and their escorts and the heroism of the pilots is remembered in both countries.

Having more mobility than the army, the remnants of the air force retreated to France and to Oujda in Morocco before Belgium abandoned the uneven fight on 27 May, and capitulated. There was insufficient time to regroup the units in France before that country, too, was forced to surrender. Thirty Belgian pilots fled to Britain where their abilities found vital employment defeating the Luftwaffe during August and September, and destroying twenty-one German aircraft.

Following the precedent established by the emigré Poles and Czechs, a Belgian element was formed in the Royal Air Force on 11 February 1941, consisting of one flight of 609 Squadron, a unit later commanded by Roland Beamont of English Electric test pilot fame. On 12 November 1941, 350 Squadron came into being as the first fully-Belgian squadron of the RAF, being joined a year later by 349 Squadron. These two units operated as a wing in the 2nd Tactical Air Force and assisted in the liberation of Europe before returning to their homeland permanently in October 1946.

During April 1945, when Russian troops were still fighting their way towards the centre of Berlin, plans for the resurrection of the Belgian Air Force were already in hand at the East Anglian airfield of Snailwell. Thirty Tiger Moths based here began the task of training pilots for the new air force, and in March 1946 transferred to Schaffen to become the Elementary Flying School (EVS). The nucleus of a transport fleet was likewise provided by the Belgian section of the Allied Communications Squadron at Hendon flying three Hurricanes, two Ansons, two Dominies, a Miles Monarch and a Stampe SV-4B.

For some time the Belgian elements remained in the RAF whilst plans were made for the re-establishment of the air force on home soil. Of the 1940 bases, only Evère was servicable with Schaffen and Wevelgem badly damaged. Four ex-Luftwaffe bases at Chièvres, Bevekom, Brustem and Florennes were additionally available, whilst Melsbroek was reserved for civilian use.

On 15 October 1946 all Belgian sections of the RAF were handed over to the new government to re-establish the Militair Vliegwezen. In addition to the EVS, 349 and 350 Squadrons left Fassberg, Germany for Bevekom on 24 October and the Technical School moved into Schaffen from Snailwell.

Initial equipment for the Luchtmacht or Force Aérien, as it later became, was from RAF stocks and involved types such as the Spitfire, Mosquito, Harvard, Oxford, Dominie, Proctor, Auster AOP.6, Anson, Martinet and Dakota (the latter officially American property).

Hardly had the last German soldier thrown down his rifle and walked into captivity, than the allies began to break up their partnership and divide into two camps, East and West. Imperceptibly at first and then with ever increasing speed, the democratic governments of Poland, Rumania, Czechoslovakia and Hungary were toppled and their place taken by Soviet 'puppets'. The process resulted in a long serpentine crack appearing in the map of Europe — almost as if some gigantic earthquake had torn the continent in half, inhibiting communication on all levels between the two parts. Winston Churchill gave this phenomenon its name, the Iron Curtain.

Hastily, Western Europe formed its own defence pact, the Western Union in Brussels on 17 March 1948 and with the later entry of the USA into this agreement, on 4 April 1949 NATO was established.

American assistance under the Mutual Development and Aid Program augmented the Meteors financed under Western Union, by Thunderjets and T-33As enabling the Spitfires to be retired. Belgium's first jet fighters had arrived at Bevekom in June 1949 when the Meteor F.4 re-equipped 350 Squadron, and in April 1951 2 Wing at Florennes took on charge its first Thunderjets. At its peak in 1955—56, the Luchtmacht contained nine squadrons of Meteor F.8, two of Meteor NF.11, ten of Thunderjets and two of transport types, principally the C-119G and C-47.

Clearly, operational squadrons were of no use without support services, and flying-training, technical-training, and navigation schools were established, together with the many other administrative and storage facilities needed. An Auxiliary Squadron flying Spitfires, and later Meteors, existed from 1949 until 1958.

The Luchtmacht went supersonic for the first time in August 1955 when Thunderstreaks arrived to replace the Thunderjets. In the next year, it was the turn of the Meteor F.8 to move-over in favour of the Hunter F.4 (and subsequently F.6) and late in 1957 the Meteor NF.11 (which had replaced the Mosquito NF.30) was supplanted by the Avro CF-100 Canuck all-weather interceptor. Apart from providing the aircraft themselves, Canada additionally trained the navigators for them at Summerside and Winnipeg.

In 1960, Belgium lost the use of a valuable training base at Kamina in the former colony of the Congo. The airfield had been opened for Belgian military aircraft in 1949 and was the terminal for supply flights of Dakotas, DC-4s and later DC-6s from the home country. During 1953 the advanced flying school, the VVS, took its Harvards to Kamina and early in 1960, Fouga Magisters began to arrive as replacements.

The first few days of July 1960 witnessed revolts in several sections of the Congo Republic's army, and as these spread, the C-119G Boxcars of 15 Wing were put on alert on the 5th. It was soon obvious that all control of the army had been lost and the lives of Europeans living in the Congo were in imminent danger. An airlift was begun by 15 Wing, carrying paratroops into defended areas to hold back the rebels until the Europeans could be rescued. They were assisted in this task by VVS Harvards and Magisters hastily armed with rockets and machine guns.

Eventually, when United Nations troops arrived, the Belgians were able to concentrate on saving their own equipment. The Magisters were freighted back to Bierset, but most Harvards were scrapped and the three Alouettes and six Doves left for the Katangan Air Force. During the campaign, Harvard H-210 was shot down together with an Alouette on 18 July and Boxcar CP-36 crashed into a mountain the next day. One further Harvard was lost.

Until the VVS could be properly reconstituted, Belgian pilots were trained by the Dutch and French air forces, but on 22 January 1962 the school opened its doors once more at Brustem. In doing so it began a programme of joint training for Dutch and Belgian pilots in which candidates from both countries received their initial jet training with the VVS, graduating then to Woensdrect in Holland for advanced training on the T-33A where, to meet the increased commitment, eight aircraft from the Belgian Air Force were transferred. Operational conversion for both nations took place at Eindhoven, Holland on the F-84F.

By 1960, the Canucks were becoming obsolete in air defence, as were the Hunters in 7 and 9 Wings, but 13 Wing had transferred to the guided missile for the defence of Belgian airspace and No.9 was soon to follow. In March of that year, Holland and Belgium agreed upon a joint requirement for 300 Starfighters of which the Belgian Air Force was to have an initial supply of 100, plus 70 more on option. 2 Wing was designated as the first recipient, but the option was cancelled and the aircraft instead went to 1 Wing in 1963 for interceptor duties, and to 10 Wing the next year to replace Thunderstreaks as ground attack aircraft. They combined with other national elements of 2ATAF; the interceptors with their Dutch counterparts as 69

Group and the fighter-bombers with the RAF's ground attack aircraft in Germany as 83 Group. Starfighters were camouflaged in 'Vietnam-type' colours during 1967.

Having had the prospective new equipment taken almost from under their noses, 2 Wing were obliged to fly their Thunderstreaks for a long time before they could be traded in for a more modern type. Despite Dutch enthusiasm for the Northrop F-5, Belgium chose the Mirage 5 in 1968 (and thus ended the common training programme) after receiving an offer from Dassault which was impossible to refuse. Eighty-five – later 103 – Mirages were built under licence by Avions Fairey and SABCA on generous terms granted by the French in order to get a toe in the door of what had been hitherto an exclusively British and American market.

Mirages eventually replaced the F-84F and RF-84F in 1970–71, just as another new type, the SIAI-Marchetti SF-260M basic trainer, supplanted Europe's last military biplane, the Stampe SV-4, at Goetsenhoven. No sooner had this been done, than it was time to look at a successor to the Starfighter and T-33/Magister for which the F-16 and Alpha Jet were ultimately chosen.

The Belgian Starfighters are expected to reach the end of their fatigue lives from early 1979 onwards and the first F-16A/B deliveries are programmed to begin in the middle of that year and be complete by 1984. Ninety F-16A and twelve F-16B twin seat trainers are on order and will re-equip 1 and 10 Wings.

Before the Alpha Jet becomes available, the Lockheed T-33 will have served almost twenty seven years with the Belgian Air Force and the Magister, eighteen. It was decided to order the Alpha Jet on 13 September 1973, but not until two years later, in September 1975, was the agreement for sixteen, plus seventeen options actually signed. First deliveries will go to 11 Smaldeel (squadron), replacing T-33As in November 1978, and following this, the balance are for 7 and 9 Smaldelen.

Not only in offensive strength has Belgium recently re-equipped its air force; the years 1970–76 have witnessed a rapid change in aircraft types, consigning some old workhorses to pasture (literally in the case of the C-119). The transport fleet has now equipped with the Hercules, Boeing 727, HS.748, Mystère 20 and Swearingen Merlin, whilst Alouette III and Sea King helicopters have modernised SAR.

Today, the Luchtmacht is under the control of Lt.Gen.Av. A. D. Debeche and employs 19,900 of all ranks, in three commands, Tactical, Training and Support. The first of these to be encountered by the student pilot is the Groepering Opleiding en Training (Instruction and Training Group) which is sub divided into flying training (Vervolmakings Centrum), general and administrative (Centrum voor Militaire Vorming) and technical (Technische School) sections.

Pilot training begins at Goetsenhoven with 125 hours of flying the SF.260MB incoporating 4 hours at night, 10 in formation, 25 on instruments and 30 in navigational exercises. From here, the Magisters of 7 and 9Smn. at Brustem (collectively the Voortgezette Vliegopleiding School – advanced flying school) provide a further 125 hours. Transitional training of 100 hours at 11Sm. (smaldeel Vliegen Zonder Zicht) precedes transfer to an operational conversion unit, either the F-104 conversion flight for the Starfighter or 8Sm. for the Mirage 5, where 32 months after joining-up, the pilot is ready for his first squadron.

Operational elements of the Luchtmacht are contained in the Groepering Tactische Luchtmacht under Maj.Gen.Av. B. E. M. De Smet and with the exception of the transport fleet of 15 Wing (20 and 21Smn.) and helicopters of 40Sm., is allocated to NATO's 2 ATAF. The group contains four Starfighter squadrons each with 22 aircraft (including 4 in reserve), two in 1 All Weather Wing and two in 10 Fighter-Bomber Wing. Two Mirage wings complete the manned offensive armament, 2 Fighter Bomber Wing with 18 Mirage 5BA of 2Sm. and 18 Mirage 5BR (Reconnaissance) of 42Sm. and 3 Tactical Wing with 1Sm. (18 Mirage 5BD) and OCU 8Sm. (having a mix of 12 Mirage 5BD trainers and six 5BA) also acting as a strike squadron in addition to its training role.

Two guided missile wings are based in Germany, the 9th and 13th, each possessing four squadrons of Nike-Hercules (16 per squadron) acting as a long stop to the Landmacht (army) Hawks. Together with a missile support wing, they form the Groepering Missiles. The third Luchtmacht command, the Basis van de Luchtmacht has no aircraft and comprises administrative and supply units only.

Landmacht

The Army received its first AOP aircraft in mid 1947, but did not take control of its own aviation section until 1 May 1954 when the base squadron at Brasschaat (15Sm.) and liaison section in Germany (16Sm.) became part of the Licht Vliegwezen. Two more units formed in 1956 flying the Super Cub, and later, Alouette II, whilst twelve Islanders replaced the Dornier 27's in 1975–76.

15 Smaldeel is now designated the 'School van het Licht Vliegwezen' and thus provides pilots and crew for the legerkorpssmaldeel (Corps Observation Squadron) and two divisional squadrons: 17Sm. attached to 16 Armoured Division and 18Sm. with 1 Infantry Division.

Zeemacht

A naval flight of two Sikorsky HSS-1 helicopters loaned from the air force at Koksijde formed in the mid sixties and was augmented in 1970–71 by three Alouette III. The Zeemacht Flight operates as part of 40 Smaldeel communicating with naval vessels at sea and providing SAR facilities.

Part Two

Units/Organisation

Unit designations

From the inception of military flying, the squadron has been the basic unit, known by its Flemish name of 'Smaldeel' or in French as 'Escadrille'. At the start of World War 2, army co-operation and bomber squadrons had a strength of approximately ten aircraft, and fighter squadrons, fifteen. Following the practise adopted by the French, there existed several squadrons with the same number, but attached to different groups e.g. 1 Bomber squadron, 1 Fighter squadron etc.

In the Army Co-operation Regiment, each squadron was attached to an individually numbered Groep (group), but in the fighter and bomber regiments, a Groep consisted of two squadrons. Individual squadron insignia was carried, but in some instances two squadrons (often in the same Groep) would have identical badges with only the colour changed for differentiation.

After World War 2, the previous system was abandoned in favour of one following more the RAF numbering sequence. At the time there were two ex-RAF squadrons, 349 and 350 in existence forming 160 Wing (RAF). As further squadrons formed, they took up RAF-type numbers, RAF-type wing designations and even imitations of RAF wartime codes. This continued until 1 February 1948 when a new system of numbering began with squadrons and wing numbers beginning at 1. The previously numbered units received new designations, but 349 and 350 Squadrons, because of their genuine wartime links, kept their old 'number-plates'.

Anticipating an increase in strength, some wing and squadron numbers were left unallocated in 1948 with the result that fighter squadrons began at 1, AOP at '15' and transport at '20'. The 'Wing' replaced the 'Groep' as the larger administrative unit, and wings usually had two or three squadrons, although 1 Wing managed six for a time.

Examination of the order of battle for 1 January 1956 (at the end of this section) will reveal several salient points on the allocation of squadron numbers. 2 Wing holds 1, 2 and 3 Squadrons, but as 4 Squadron went to 1 Wing, the next unit to form — 7 Wing — took up 7, 8 and 9 Squadrons. A change in policy is reflected in 9, 10, the abortive 5, and 13 Wings, where the first squadrons in each were 22–25 Squadrons, the second 26–29 Squadrons, and the third 30–33 Squadrons, a method of allocation similar to the dealing of playing-cards, resulting in some units, e.g. 28 and 32 Smaldelen, never having been formed, when the establishment of 5 Wing was curtailed.

Squadron strengths during the early Fifties were normally 25 aircraft for the fighter units, but as attrition took its toll, some wings reduced to two squadrons to maintain full Smaldeel complements. In the post-war period, old unit insignia were resurrected, but could not often be given to a squadron with a similar number or even function.

The use of three-letter codes for fighter aircraft was discontinued early in the sixties, but in 1971 the Luchtmacht adopted another RAF practise in allocating 'Shadow Squadron' numbers to training units, with the result that the advanced flying school at Brustem was divided into 7 and 9 Smaldelen, and the blind-flying establishment there became 11 Smaldeel.

Reduced expenditure on defence now dictates that all but the missile wings have only two attached squadrons.

Unit histories

In the following pages are the histories of Luchtmacht units formed since 1946, including those having no aircraft.

The Belgian practise of allocating the same badge to units of a different designation, as in the French Air Force, causes confusion in the tracing of squadron histories. Here, the insignia is used as the basis of the history, and all units having used that particular badge are listed beneath. Any insignia which was not perpetuated after World War 2 has not been illustrated.

Identities are likewise allocated to Wings (usually comprising three squadrons at inception, but now normally having two) with the result that it has been necessary to duplicate parts of some squadron histories in the section dealing with Wings. Readers conducting research are advised to refer to both Smaldeel and Wing histories as this duplication has been kept to a minimum. Order of battle tables at the end of this section will assist the cross-checking of unit equipment, locations and parent formations.

Units are listed in the sequence: (1) Wing histories, (2) Smaldeel histories, and (3) Unnumbered units.

Those units currently extant are readily traced by reference to the 'status' side heading.

1 WING ALLE-WEDER

Insignia: A golden falcon on a blue disc. Lettering in gold. Used in lieu of squadron insignia.
Motto: Audeo Aciem (I dare the battle)
Status: Currently comprises No's 349 and 350 Smaldelen at Bevekom, equipped with TF- and F-104G Starfighters.

1 Jachtwing formed on 1 February 1948 out of 160 Wing (349 and 350 Squadrons) flying Spitfire

XIVs at Bevekom and quickly added 10 Smaldeel, with Mosquito night-fighters. With the arrival of the Meteor F.4 in 1949, 4 Smaldeel was added and in 1951, this unit, plus 349 and 350 Smaldelen re-equipped with the Meteor F.8 - being the second Wing to do so. At the same time a second Mosquito Smaldeel (No. 11) was established and an Auxiliary Squadron (Hulp Smaldeel) of Spitfires (later with Meteor F.4s) was attached. 11 Smaldeel received the Meteor NF.11 in July 1952 and 10 Smaldeel followed suit in 1956.

With the arrival of the Hunter F.4 in 1957, 4 and 10 Smaldelen disbanded, but within a year, 11, 349 and 350 Smaldelen converted to the CF-100. 11 Smaldeel disbanded in 1960, followed by 350 early in 1963, but the latter received Starfighters in the April of that year, as did 349 Smaldeel, soon afterwards.

From 1960 to 1963, the Thunderflashes of No. 42 Verkenningssmaldeel were attached to the Wing prior to transfer to 3 Wing at Bierset.

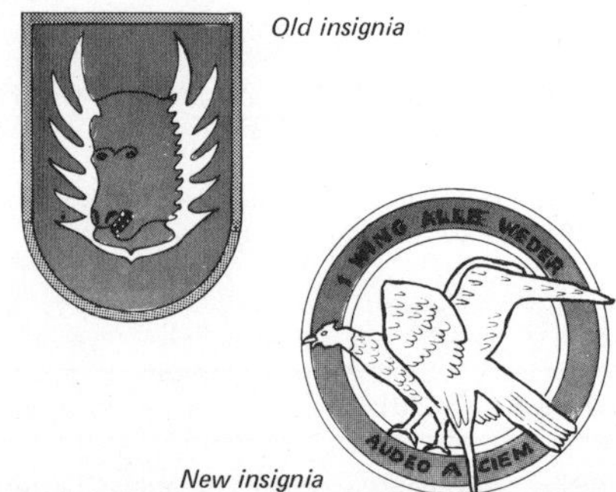

Old insignia

New insignia

Prior to 1965, the Wing emblem/insignia was of a light blue wolf's head with red mouth and white teeth, flanked by a pair of white wings and mounted on a blue shield with red borders.

The two squadrons presently fly the F-104G in the all-weather interceptor role and will probably receive the F-16 in due course.

2 TACTISCHE WING

Insignia: A golden boar's head and golden hunting horn on a black shield with green border.
Motto: Taiaut-Taiaut
Status: Currently comprises No's 2 and 42 Smaldelen at Florennes, equipped with the Mirage 5BA and 5BR.

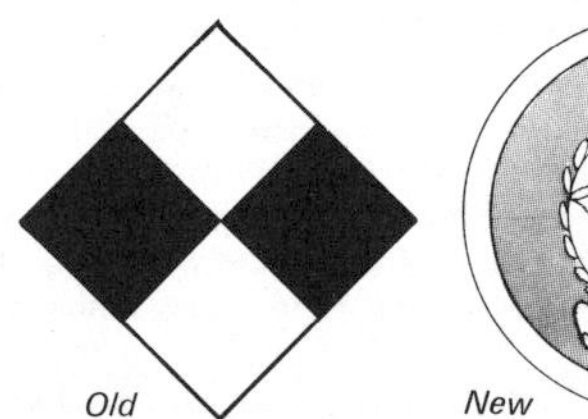

Old

New

2 Wing was formed on 1 February 1948 by renumbering 161 Wing, which comprised 1 and 2 Smaldelen both flying Spitfire XIVs. 3 Smaldeel was added in the summer of 1949, equipping with the Spitfires released when 349 and 350 Smaldelen received their Meteors. Thunderjets arrived in April 1951, prompting a change of role from fighter to fighter-bomber duties and these were supplanted by Thunderflashes in August 1955, commencing with 3 Smaldeel. In August 1956 Florennes was named *J. Offenberg* base and 2 Smaldeel was disbanded on 15 October 1960.

The OCU for the Mirage 5 (8 Smaldeel) formed on 15 July 1970, following which, 2 Smaldeel received the new type. On 2 July 1971 the Thunderstreak-equipped 1 Smaldeel moved to Bierset, making room for 42 Smaldeel to arrive with its first few Mirages on 15 September.

With the advent of the Mirage, 2 Wing changed its function from Jagers-Bommenwerperswing to Tactische Wing, and despatched 8 Smaldeel to Bierset on 15 December 1971 to join 3 Wing.

Aircraft of 2 Wing were marked with a black and white quartered diamond in place of the boar.

3 TACTISCHE WING

Insignia: A brown-grey bison on a blue shield with red borders. Golden lettering.
Motto: Ardor et Valor (Zeal and courage)
Status: Currently comprises No's 1 and 8 Smaldelen at Bierset, equipped with Mirage 5BA and 5BD versions.

3 Wing was formed on 4 December 1968 from two Bierset-based units: the Smaldeel Vliegen Zonder Zicht (formed 1962) and 42 Verkenningssmaldeel (resident since 1963). The Wing adopted the bison insignia of 26 Smaldeel, 9 Wing, a previous base resident, the following year.

In May 1971 the sVZZ departed to Brustem, making way for 1 Smaldeel from Florennes on 2 July, with its Thunderstreaks. After recieving a few Mirage 5BR aircraft, 42 Smaldeel left for Florennes on 15 September to complete its re-equipment, donating its remaining Thunderflashes to 1 Smaldeel. The Mirage OCU (8 Smaldeel) took up residence on 15 December 1971, arriving from Florennes with its Mirage 5BA and 5BDs, to train the pilots of 1 Smaldeel when their turn came to receive the new type in 1972.

4 WING

Status: Not formed

5 WING

Insignia: Nil
Motto: Nil
Status: Not currently formed

Intended as the fourth Meteor F.8 Wing, the unit was established late in 1952 with the formation of its first squadron, 24 Smaldeel. A change of policy resulted in the non-activation of its intended other constituents (28 and 32 Smaldelen) and instead, 24 Smaldeel acted as a target-towing squadron for the Jachtschool - its sister unit at Koksijde, and for other BLu Smaldelen, at the Sylt ranges in Northern Germany. The identity of 5 Wing was lost in 1954 when 24 Smaldeel became the 'Sleepvlucht' or Target Towing Flight. The last aircraft type used was the Meteor F.8 at Koksijde.

6 WING

Status: Not formed

7 WING

Insignia: Three folded paper cygnets, in red, blue and green on a silver shield.

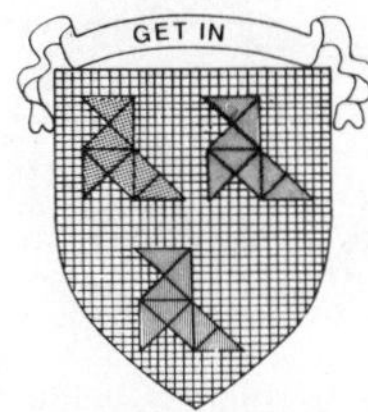

Motto: Get in
Status: Not currently formed

7 Wing formed at Chièvres on 1 December 1950, under Major Van Lierde, but it did not receive its first squadron (7 Smaldeel, with Meteors) until 1 March 1951. In the June of that year 8 Smaldeel formed and this was followed twelve months later by 9 Smaldeel. All three constituent squadrons converted to the Hunter in mid-1956; 9 Smaldeel disbanded in March 1957, afterwhich the wing received Hunter F.Mk.6s - the F.Mk.4s being passed on to 1 Wing. The 7th retained Hunter F.6s until November 1963, when its two squadrons disbanded at Chièvres.

8 WING

Status: Not formed

9 DAGJACHT WING

Insignia: Nil
Motto: Nil
Status: Not currently formed

9 Jagers-Bommenwerperswing was formed at Bierset on 2 September 1953 as the Luchtmacht's third F-84G Thunderjet wing. Its first component squadron, (22 Smaldeel) was commissioned the same day; 26 Smaldeel following in July 1954 and 30 Smaldeel in October 1955.

A change from fighter-bomber to day-fighter duties was reflected in the redesignation to 9 Dagjachtwing on 1 July 1956 and the adoption by 22 and 26 Smaldelen of the Meteor F.8, whereas 30 Smaldeel disbanded in October 1956.

The Meteors were soon replaced by the Hunter F.6 and this aircraft continued to serve 9 Wing until both the component squadrons were disbanded at Bierset on 30th April 1960.

Just two years later, 9 Wing re-emerged, but as a missile unit.

9 WING MISSILES

Insignia: In a divided shield of light blue and dark blue, a white Nike-Hercules missile emitting red flames; also a black figure nine.
Motto: Nil
Status: Currently comprises No's 53, 54, 55 and 56 Smaldelen administered from Grefrath and equipped with Nike-Hercules missiles.

Technical personnel of Belgium's second missile wing received training at the U.S. Missile Training Centre, Fort Bliss in 1961; returning home to Bierset where 9 Wing was established on 17 February 1962 with 54 and 55 Smaldelen component units. The Wing became operational under the title '9 Wing Telegeleide Tuigen Grond-Lucht' on 28 May 1962, but the title was later shortened to '9 Wing Missiles'.

9 Wing headquarters has been at Grefrath since 1969, from where are administered 53 Smaldeel (ex 13 Wing on 12 August 1964) at Kaster; 54 Smaldeel; 55 Smaldeel at Kapellen-Erft and 56 Smaldeel (formed November 1953) at Grefrath.

10 BOMMENWERPERS WING

Insignia: A red lion and golden crown upon a blue shield. Lettering in red (used in place of squadron insignia).

Motto: Fortuna Favet Fortibus (Fortune follows the brave)
Status: Currently comprises No's 23 and 31 Smaldelen, based at Kleine Brogel, equipped with TF- and F-104G Starfighters.

On 20 December 1951, 23, 27 and 31 Smaldelen moved from Bevekom to Chièvres where they combined to form 10 Wing, with Spitfire XIVs. The next year saw the arrival of Thunderjets, with 31 Smaldeel also taking on charge a few T-33A advanced instrument trainers to assist in its task of blind-flying instruction.

10 Jagers-Bommenwerperswing moved to Bruggen on 20 March 1953, remaining there until Kleine Brogel was completed in 1954, though detachments were made to Bierset, Geilenkirchen and Sylt.

The first Thunderstreaks arrived in June 1956 and these were to serve until the Spring of 1964 when F-104G Starfighters were received for 23 and 31 Smaldelen - 27 Smaldeel having disbanded in June 1962.

11 WING

Status: Not formed

12 WING

Status: Not formed

13 JACHTWING

Insignia: Nil
Motto: Nil
Status: Not currently formed

Formed at Koksijde in December 1953, 13 Wing was the Luchtmacht's third Meteor F.8 wing and comprised 25, 29 and 33 Smaldelen. The Wing moved to Brustem in 1954 and 33 Smaldeel disbanded 2 May 1957 following a directive that the Meteor wings be restructured on a two-squadron basis. The retirement of the Meteor F.8 was followed by the disbandment of the remaining 25 and 29 Smaldelen in July 1958, though 13 Wing reformed the next year with Nike missiles.

13 WING MISSILES

Insignia: In a divided shield of light blue and dark blue, a white Nike-Hercules missile emitting red flames; also the figures '13' in black.
Motto: Nil
Status: Currently comprises No's 50, 51 and 52 Smaldelen administered from Düren, and equipped with Nike- Hercules missiles. 57 Smaldeel is expected to form and become part of the Wing during 1977.

In May 1959, the first four missile squadrons (50, 51, 52 and 53 Smaldelen) formed in 13 Wing, which was then based at Lombardsijde, though later moved to Elsenborn and ultimately to Bodart barracks, Düren.

53 Smaldeel transferred to 9 Wing on 12 August 1964, in return for which 54 Smaldeel was attached on 25 October. The components are thus: 50 Smaldeel at Düren, 51 Smaldeel at Blankenheim and 52 Smaldeel at Euskirchen.

14 WING

Status: Not formed

15 VERVOER EN VERBINDUNGSWING

Insignia: A shield outlined in red within which is a Sioux' head. The head-dress has black and white feathers and a white headband with black triangles over the forehead, and is attached to the white hair by a blue cockade. The head is contained within a light blue disc with dark blue and red outlines.
Motto: Tenacité (Tenacity)
Status: Currently comprises No's 20 and 21 Smaldelen, operating Hercules, Merlin, HS.748 Mystère 20 and Boeing 727 aircraft.

15 Vervoer en Verbindungswing (Transport and Communications Wing) formed at Evère 1 February 1948 out of 169 Wing and comprised 20 and

21 Smaldelen, specialising in transport and communications respectively. The Wing moved to Groenveld Camp, Melsbroek in 1950, where 21 Smaldeel received the Pembrokes, DC-4 and DC-6 (the latter two for communications with the Congo) and 20 Smaldeel received the C-119G Packet. A second C-119G squadron, 40 Smaldeel, existed for a few months in 1954, only to reform in June 1961 with the same aircraft, which it flew until they were withdrawn in 1973. At the same time 20 Smaldeel converted to the Hercules and now 21 Smaldeel operate a selection of types as listed above. As 15 Wing is resident on the military side of Brussels/Zaventem Airport, its aircraft are conveniently placed for servicing by the airline SABENA.

21 LOGISTIEKE WING

Status: Currently responsible for the control of aviation supplies and weapons, dealing with government and civilian sources. As a secondary role it seeks to improve organisation by adopting modern methods and eliminates faults and weaknesses in equipment, especially where it may endanger life.

The Wing consists of four groups:— **Groep Ramingen en Identificatie:** estimation of requirements and standardisation of supplies and material. **Groep Stocks:** reception, storage, stock-control and allocation of supplies. **Groep Werkplaatsen:** repair, maintenance and production of materials, transportation of heavy supplies. **Basis Groep:** day-to-day administration of 21 Wing.

22 LOGISTIEKE WING

Status: Currently responsible for the supply of all electronic material to the air force, including guided missiles, at Evère.

The origins of the Wing can be traced back to the formation of the army's No.5 Compagnie Radio-Telegrafie, the 'Luchtvaartcompagnie' in 1936. This was enlarged as the 'Bataljon TTR-Luchtvaart' in 1939 with an HQ staff and one Company attached to each of the three Air Regiments.

During 1947, Belgian specialists assumed responsibility for telecommunications from the RAF units previously employed and on 1 April 1951 this was organised as the OnderhoudsEenheid van het Radar Scherm (Radar Defence Supply Unit) comprising a groep Radio-Radar and a groep Telefonie-Telegrafie, later adding an Onderhoudsgroep (Supply group).

The OERS was re-organised on 3 March 1954 as the Unité de Télécommunication, whereupon the Radio-Radar group and Onderhoudsgroep were brought together as the Groep Werkhuizen (Workshop group). At the same time, the groep Telefonie-Telegrafie became the Groep Verbindingen (Communications group) and a Basis Groep (HQ) established, the former providing all installations for all Luchtmacht ground units.

On 1 July 1959 a reorganisation of military supply units allocated to the OERS sections two new groups, the Groep Stocks and the Groep Ramingen en Identificaties (Estimation and Identification group) whilst the Groep Verbindingen became autonomous as the Groep (later Wing) Telecommunicaties. The four remaining sections were designated No. 2 Depot and on 1 October 1966, 22 Logistieke Wing.

The Wing consists of four groups:— **Groep Ramingen en Identificatie:** Estimation of requirements and standardisation of supplies. **Groep Stocks:** Holding of supplies to the Groep Werkhuizen. **Groep Werkuizen:** Repair and overhaul, installation of new and repaired equipment, contracts with civilian firms. Groep includes sektie Metrologie, a measurement and performance testing section. **Depot voor Fotografisch Materieel:** Photographic and camera store.

23 LOGISTIEKE WING

Status: Currently responsible for the supply and maintenance of vehicles, ground equipment, handling equipment, clothing and kit, chemicals, petrol. tools and household articles. The unit is stationed at Eckstein Barracks, Zellik, and was established in 1950 out of Materieel Bevoorradingsdepot 3.

The Wing consists of four groups:— **Groep Ramingen en Identificatie:** Forward planning group. **Basis Groep:** Administration group. **Groep Werkplaatsen:** Repair and overhaul. **Groep Stocks:** Storage group.

24 WING

Status: Not formed

25 LOGISTIEKE WING

Status: Currently handles the supply of all explosives used by the air force.

The first such establishment formed at Leuven-Namen in 1925 and existed until taken over by the Germans in 1940. It was later re-established as 'Depot No.5' and renamed 25 Wing in 1965.

The Wing consists of four groups:— **Groep Stocks:** Storage group. **Groep Ramingen, Identificatie en Codificatie:** Handles requisitions from units and internal organisation of storage areas. **Groep Munitiewerkplaatsen:** Responsible for munitions depots and works. **Dienst voor Opruiming en Vernietiging:** bomb disposal and safety unit also responsible for investigating aircraft accidents involving weapons malfunctioning.

160 WING

Insignia: Nil
Motto: Nil
Status: Not currently formed

With the end of hostilities in Europe, 123 Wing at Wunsdorf, comprising 349 and 350 Squadrons, moved to Fassberg as 160 Wing, remaining there until returned to Belgian control on 15 October 1946. Nine days later, the Wing left Fassberg to become part of the Belgian Luchtmacht and retained its former title until 1 February 1948 when it was re-numbered 1 Wing at Bevekom, flying the Spitfire XIV.

161 WING

Insignia: Nil
Motto: Nil
Status: Not currently formed

On 1 November 1946, 351 Squadron formed as Belgium's third Spitfire unit and the first in 161 Wing. It was joined soon afterwards by 352 Squadron, also with Spitfires, this being re-numbered 2 Jagers-Bommenwerperssmaldeel on 1 November 1947, with 351 Squadron likewise becoming 1 Jagers-Bommenwerperssmaldeel on 10 January 1948. On 1 February the wing was re-designated 2 Wing, at its Florennes base.

169 WING

Insignia: Nil
Motto: Nil
Status: Not currently formed

169 Wing was Belgium's first transport wing and formed on 1 April 1947 at Evère with 366 and 367 Squadrons (neither of which were RAF numbers).

Crews for 169 Wing came principally from the Belgian crews of two RAF squadrons equipped with Dakotas.

The first of these, 187 Squadron, was engaged on transport flights from Britain to bases in Belgium and northern Germany and the second, 525 Squadron, delivered mail and newspapers to the armies of occupation. Both were based at Memburg in 1946 but returned to England in the latter part of the year where they were re-numbered.

169 Wing also absorbed pilots from the RAF Fighter, Bomber and Coastal Commands, and immediately upon establishment, commenced regular flights to Hendon, Prestwick, Villacoublay and Wahn. The two component Squadrons, 366 and 367, had a strength of five Dakotas, nine Ansons, six Oxfords, four Dominies, one Magister and three Proctors at the time of formation. The unit was re-numbered 15 Wing on 1 February 1948 and its two squadrons likewise became 20 and 21 Smaldelen.

1 SMALDEEL

Insignia: A green thistle surrounded by a belt.

Motto: Nemo Me Impune Lacessit
(No man provokes me with impunity)
Status: Currently a component of 3 Wing based at Bierset and equipped with Mirage 5BA

The origins of 1 Smaldeel may be traced to the genesis of military flying in Belgium when the 'Compagnie van Werklui en Luchtschippers' (Aircraft Crew and Support Corps) was formed in 1910 under Commandant Le Clément de Saint-Marcq. A royal command of 16 April 1913 established the 'Compagnie van Vliegers' (Flying Corps) of which 1 Smaldeel at Brasschaat was the first component, flying 16 Farmans and commanded by Lt. Demanet.

At the beginning of World War 1, the squadron was stationed at Ans and attached to 3 Division of the army, moving later to Oostende and St. Pol as the Germans advanced. A re-organisation of the air force under the title Militaire Luchtvaart in March 1915 brought 1 Smaldeel to St. Idesbald and later, Moeres where in 1917 the squadron insignia of a Scots thistle was designed by André de Meulemeester.

Following the German pattern, Fighter Groups (Groepering Jacht) were established in February 1918, and 1 Smaldeel became 9 Smaldeel, paired with 10 Smaldeel (ex No. 5), and 11. Following the Armistice 9 Smaldeel moved to Schaffen, remaining there, and by 1940 equipped with Hurricanes under the old 1 Smaldeel designation within 1 Group, 2 Regiment. On the first day of the German attack, 10 May 1940, nine of the eleven Hurricanes were destroyed on the ground and the squadron played no further part in the battle.

The thistle emblem re-appeared on 1 November 1946 on the Spitfires of 351 Squadron, 161 Wing (codes '3R- ') at Florennes, this being re-designated 1 Smaldeel on 10 January 1948. The Spitfire XIV gave way to the F-84E/G Thunderjet, the first of which arrived on 18 April 1951; the F-84F Thunderstreak in August 1955 and Mirage 5BA in 1972. Prior to receiving its new equipment, 1 Smaldeel transferred to Bierset on 2 July 1971 where it additionally flew ex-42 Smaldeel Thunderflashes on photo reconnaisance missions until the latter was fully operational. 1 and 8 Smaldelen now form 3 Wing.

2 SMALDEEL

Insignia: A red comet of five points and a tail
Motto: Ut Fulgur Sulca Aethera
(Cleave the air like lightning)
Status: Currently a component of 2 Wing based at Florennes, flying Mirage 5BA.

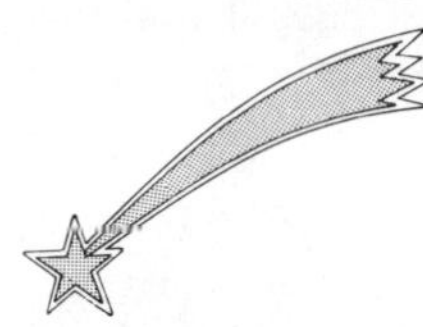

The comet insignia employed today by 2 Smaldeel originated in 1917 on the aircraft of 5 Jachtsmaldeel under Commandant Jacquet. In the next year, the unit was re-designated 10 Smaldeel and in 1926 became 1 Jachtsmaldeel of 1 Groep. On the outbreak of World War 2, 1 Smaldeel was equipped with Gloster Gladiators but soon lost these obsolete fighters in the face of the Luftwaffe onslaught in May 1940.

On 1 November 1947, 352 Squadron in 161 Wing was re-numbered 2 Jagers-Bommenwerperssmaldeel, equipped with the Spitfire XIV at Florennes (coded 'VR- '). 161 Wing, likewise became 2 Wing on 1 February 1948 and the two units have remained in association, flying the Thunderjet, Thunderflash and, since 1971, the Mirage 5BA.

3 SMALDEEL

See 52 Smaldeel

4 SMALDEEL

See 51 Smaldeel

5 SMALDEEL

Status: Not formed

6 SMALDEEL

Status: Not formed

7 SMALDEEL

Insignia: A silver cygnet made of folded paper on a red background.
Motto: Nil
Status: Currently a 'shadow squadron' at Brustem, equipped with CM.170 Magisters.

The origin of the cygnet insignia is explained under the section dealing with 56 Smaldeel. When, however, 7 Wing formed on 1 December 1950 it allocated a different coloured version of this motif to each of its components as they formed.

7 Dagjachtsmaldeel (day fighter squadron) was established at Chièvres on 1 March 1951 equipped with the first Belgian Meteor F.8s (coded '7J- '). Hunter F.4s were received in 1956 and the F.6s arrived just a year later. The squadron disbanded in November 1963, but was re-born in June 1971 when one half of the Magisters of the 'VVS' received the appelation 7 Smaldeel in the manner of a 'shadow squadron'.

8 SMALDEEL

Insignia: A silver cygnet made of folded paper on a blue background. (See 56 Smaldeel)
Motto: Nil
Status: Currently a component of 3 Wing based at Bierset and equipped with Mirage 5BD and Mirage 5BAs, in a dual ground attack and training role.

8 Dagjachtsmaldeel formed at Chièvres in 7 Wing in June 1951 flying Meteor F.8s (coded 'OV- '). Together with the remainder of the Wing 8 Smaldeel received the Hunter F.4 in 1956, Hunter F.6 in 1957 before disbanding in November 1963.

With the arrival of the Mirage in 1970, 8 Smaldeel re-formed at Florennes on 15 July with two French-built aircraft and a nucleus of eight pilots fresh from a training course with ECT2/2 at Dijon. The unit was officially constituted on 1 August and commenced training duties on 5 October soon receiving its full complement of twelve Mirage 5BD and six Mirage 5BA. The squadron departed 2 Wing on 15 December 1971 and took up residence at Bierset under 3 Wing with whom it remains, combining instructional duties with its dual role as a ground attack squadron.

9 SMALDEEL

Insignia: A silver cygnet made of folded paper on a green background. (see 56 Smaldeel)
Motto: Nil
Status: Currently equipped with CM.170 Magisters at Brustem in a 'shadow squadron' role.

The third squadron of Meteor F.8s in 7 Wing was 9 Dagjachtsmaldeel, formed in June 1952 at Chièvres with code letters 'S2- '. Following re-equipment with the Hunter F.4, 9 Smaldeel disbanded on 15 March 1957 leaving 7 and 8 Smaldeel to continue in service.

In June 1971 the Magisters of the VVS were split into two squadrons, one of these being 9 Smaldeel.

10 SMALDEEL

See 53 Smaldeel

11 SMALDEEL

Insignia: A grey bat with black wings in a yellow triangle.
Motto: Nil
Status: Currently equipped with T-33As and based at Brustem.

As exemplified by its insignia, 11 Smaldeel was a night fighter squadron, formed at Bevekom with the Mosquito NF.30 (codes 'ND- ') in 1951 as part of 1 Wing. In July 1952, 11 Smaldeel received its first Meteor NF.11s, losing these in favour of the CF-100 in 1958, and finally disbanding during 1960.

In May 1971 the Instrument Squadron, the sVZZ, moved to Brustem to join the Vervolmakings Centrum and took up the identity 11 Smaldeel a month later.

12 SMALDEEL

Status: Not formed

13 SMALDEEL

Status: Not formed

14 SMALDEEL

Status: Not formed

15 SMALDEEL - SCHOOL VAN HET LICHT VLIEGWEZEN

Insignia: A bee
Motto: Semper Labora (Always strive)
Status: Currently based at Brasschaat, equipped with FBN Islander, Alouette II and the Alouette Astazou.

The insignia of 15 Smaldeel is derived from that of 6 Verkenningssmaldeel, established at Houtem-bij-Veurne in February 1916. Initial equipment was the BE.2c, which sounding like the Flemish word 'Bij' when pronounced, resulted in the bee marking appearing on the squadron aircraft. After World War 1, 6 Smaldeel moved to Evère and in 1924 was placed under III Groep of I Verkenningsgroepering at Goetsenhoven, disbanding in 1926 when the air force was re-organised.

On 31 July 1947, 369 Smaldeel was formed with the Auster at Braaschaat becoming 15 Smaldeel on 1 February 1948 and taking on the insignia and traditions of 6 Smaldeel on 1 September 1949. The Piper Super Cub replaced the Auster in 1955, 15 Smaldeel having passed from air force to army administrative control on 1 May the previous year.

Braaschaat is the home of Belgian army aviation and 15 Smaldeel naturally took over pilot and observer training from the air force when other army squadrons formed. In 1964 it was re-named the Schoolsmaldeel and in 1972, the School van het Licht Vliegwezen (School of Light Aviation) with twenty four Alouette II and six Do 27 on strength (the latter to be replaced by Islanders). Mountain flying-training is undertaken at Saillagouse by the French Army.

16 SMALDEEL

Insignia: A bird perched on a branch, in front of a crown and outstretched wings.
Motto: Nil
Status: Currently based at Butzweilerhof, equipped with FBN Islander, Alouette II and Alouette Astazou.

In June 1951 two Austers detached from 15 Smaldeel took up residence at Wahn in support of the Belgian army units stationed in Germany. On 1 September 1953 the flight moved to Butzweilerhof where it became 16 Smaldeel AOP, transferring to army control on 1 May 1954 as 16 Smaldeel Licht Vliegwezen.

The first three Alouette II helicopters were delivered to 16 Smaldeel in 1959 and stationed in Rwanda-Burundi from May 1960 until it passed on to the Congolese AF.

A re-organisation of army aviation in 1964 resulted in 15 Smaldeel becoming a school squadron and 16 Smaldeel a legerkorpssmaldeel (Army Corps Liaison Squadron) equipped with the Do 27. As a result its previous attachment to the 16 Pantserdevisie (Armoured Division) was taken over by 17 Smaldeel. Until 1976, 16 Smaldeel operated six Do 27, but these have now been replaced by the FBN Islander.

17 SMALDEEL

Insignia: On a shield, two scrolls in saltaire, flanked by two wings.
Motto: Nil
Status: Currently based at Werl, equipped with Alouette II and Alouette Astazou.

17 Smaldeel formed in June 1956 at its present base as an observation and liason unit for 1 Infanteriedivisie originally equipped with Super Cubs. These it followed with the Alouette II and Alouette Astazou.

Following the 1964 re-organisation of army squadrons, 17 Smaldeel took over the support of 16 Pantserdivisie stationed in Germany.

The unit additionally has a demonstration flight, the 'Blue Bees' with Alouette II helicopters.

18 SMALDEEL

Insignia: A bee, holding a pennant
Motto: Nil
Status: Currently based at Merzbruck, equipped with Alouette II and Alouette Astazou.

Established at Merzbruck in July 1956, 18 Smaldeel was originally a legerkorpssmaldeel (Army Corps Squadron) equipped with the Piper Super Cub. In 1964 its role was passed to 16 Smaldeel and the squadron instead took responsibility for liaison and observation for 1 Infanteriedivisie with its recently acquired Alouettes.

19 SMALDEEL

Status: Not formed

20 SMALDEEL

Insignia: A blue Sioux' head with black and white eagle feathers tipped with blue. A white forehead band has black triangles on it and a red disc over the ear. Black hair. Gold lettering on blue circle.
Motto: Tenacité (Tenacity)
Status: Currently based at Melsbroek as part of 15 Wing and equipped with Hercules.

In 1935, 1 and 3 Bommenwerperssmaldelen disbanded and were replaced by 9 and 11 Verkenningssmaldelen, forming V Groep of 1 Regiment. Both flew the Renard R31 and adopted the Sioux badge, that of 9 Smaldeel being circled in blue, to replace the elephant insignia of 1 and 3 Smaldelen.

The re-establishment of a transport wing in April 1946 brought into being 366 Squadron of 169 Wing at Evère equipped with Ansons and Dakotas. On 1 February 1948 the unit became 20 Transportsmaldeel, adopting the Sioux head in a blue circle as part of 15 Wing (call-signs OTCNA-OTCNZ) and moving to Melsbroek in 1950. Here it received the C-119 in 1952, operating a conversion flight with Boxcars coded '1Q- ' for a short while. The C-119 Boxcar flew continuously with 20 Smaldeel for twenty years until replaced by the C-130H Hercules in 1972.

21 SMALDEEL

Insignia: A Sioux' head, as 20 Smaldeel, but circled in red.
Motto: Tenacité (Tenacity)
Status: Currently a component of 15 Wing, based at Melsbroek and equipped with Merlins, HS.748s, Mystere 20 and Boeing 727.

The red-circle Sioux emblem was adopted by 11 Verkenningssmaldeel, V Groep, 1 Regiment on its formation at Bierset in 1935. On mobilisation in 1939, 11 Verkenningssmaldeel established its own Groep, the VI, and it was possibly one of the Renard R31s of this squadron which made the last Belgian reconnaissance flight of the German invasion forces at 1630 hours on 27 May 1940.

When peace returned, 367 Transport Squadron of 169 Wing at Evère, flying Dakotas, Dominies and Oxfords was established, becoming 21 Verbindingssmaldeel (Communications Squadron) of 15 Wing on 1 February 1948 with call-signs OTCWA-OTCWZ. 21 Smaldeel accompanied the Wing on its move to Melsbroek in 1950 and in the same year received the first of two DC-4s. These were employed on the regular run to Kamina in the Congo cutting the 4-5 day Dakota flights inaugurated in 1948 to one and a half days, and all the Harvards, Sycamores and later Magisters used there were flown-in as freight by 21 Smaldeel.

In 1952 the arrival of the C-119F with 20 Smaldeel provided the 21st with the surplus Dakotas of the former, and twelve Pembrokes arrived to replace the Ansons in 1954. Some Pembrokes had glazed noses for aerial photography by a unit attached to 15 Wing, the Exploitatie en Studiecentrum voor Fotografie (Photographic Operations and Investigation Centre).

An expansion of the long range transport fleet occurred in 1958 with the delivery of four DC-6 aircraft which joined the '4s' on flights to Kamina and Rwanda-Brundi. However, by the early 'Seventies newer equipment was required, and 21 Smaldeel took delivery of small numbers of Boeing 727s, HS748s, Mystère 20s and Merlins, the latter replacing Pembrokes in 1976 and being used as required by the photographic school.

22 SMALDEEL

See 54 Smaldeel

23 SMALDEEL

Insignia: A red devil holding two black bombs edged in yellow, above yellow flames.
Motto: Diabolico Furore (With devilish fury)
Status: Currently operates F-104G Starfighters in the fighter-bomber role as part of 10 Wing, and based at Kleine Brogel.

23 Smaldeel formed on 1 September 1951 at Bevekom, equipped with the Spitfire XIV, (codes 'Z6- '), and joined 27 and 31 Smaldelen in 10 Wing at Chièvres on 20 December 1951. Re-equipped with F-84G Thunderjets the Wing re-deployed to Kleine Brogel on 20 March 1953, later receiving the Thunderflash, and in early 1964, the F-104G Starfighter which it operates in the fighter-bomber role.

24 SMALDEEL

Insignia: Nil
Motto: Nil
Status: Not currently formed

24 Smaldeel formed with the Meteor F.8 at Koksijde in 1952 as the intended first squadron of three in 5 Wing. The aircraft (coded 'XO- '), were instead used for target-towing under the title Target Towing Flight and operated detachments to Sylt and later Decimomannu. The squadron was re-named the Sleepvlucht in 1954.

25 SMALDEEL

See 50 Smaldeel

26 SMALDEEL

Insignia: A silver bison on a red disc. Lettering in silver.
Motto: Dente Cornuque Pugnat
(He fights with tooth and horn)
Status: Not currently formed

The second squadron of 9 Wing formed in July 1954 with Thunderjets (codes 'JE- '), transferring to the Meteor F.8 on 1 July 1956 as a fighter squadron, and later the Hunter F.6 for interception duties. 9 Wing disbanded at Bierset on 30 April 1960 and the bison emblem adopted by 3 Wing.

27 SMALDEEL

Insignia: A black panther with red eyes, on a brown rock with golden edges, all within a blue shield with white borders.
Motto: Nil

Status: Not currently formed

27 Jagers-Bommenwerperssmaldeel formed on 1 September 1951 at Bevekom employing codes 'RA- '. Together with 23 and 31 Smaldelen it moved to Chièvres on 20 December 1951 where it later discarded Spitfires in favour of the F-84G Thunderjet. The Wing transferred to Kleine Brogel on 20 March 1953 converting to F-84F Thunderflashes in 1956, 27 Smaldeel acting as an OCU until disbanded in June 1962.

28 SMALDEEL

Intended as the second Meteor F.8 squadron of 5 Wing, but not formed.

29 SMALDEEL

Insignia: A black and white checkered board and two green wings all on a blue disc, surrounded by a yellow circle.
Motto: Custodiento Audere
(To dare, by guarding)
Status: Not currently formed

29 Smaldeel was formed at Koksijde in December 1953 alongside 25 and 33 Smaldelen within 13 Wing, and flew the Meteor F.8 at Brustem (codes 'MS- '), until disbanded on 1 July 1958.

30 SMALDEEL

Insignia: The brown head of a panther with red mouth, white background, yellow circle and black lettering.
Motto: Oderint dum Metuant
(Let them hate, provided they fear)
Status: Not currently formed

The third squadron of 9 Jachtwing formed in October 1955 using aircraft coded 'EB- '. When the Wing changed responsibilities from day fighter to tactical fighter and re-equipped with the Hunter F.6, 30 Smaldeel disbanded at its base at Bierset in October 1956, whilst still equipped with the original F-84G Thunderjets.

31 SMALDEEL

Insignia: A black and gold tiger

Motto: In Sanguine Vinum (Wine in the blood)

Status: Currently a component of 10 Wing, based at Kleine Brogel and equipped with TF- and F-104G Starfighters.

31 Smaldeel formed at Bevekom on 1 September 1951 flying the Spitfire XIV, (codes '8S- '). It was incorporated into 10 Wing on 20 December at Chièvres and converted to Thunderjets the next year. A move was made to Kleine Brogel on 20 March 1953, although the squadron was detached to Germany for some time. 31 Smaldeel received Thunderstreaks in 1956, operating as an OCU with fifteen aircraft on charge, and adopted its present type, the Starfighter, in mid-1964.

32 SMALDEEL

Intended as the third Meteor F.8 squadron of 5 Wing, but not formed

33 SMALDEEL

See 55 Smaldeel

34 – 39 SMALDELEN

Status: Not formed

40 TRANSPORTSMALDEEL

Insignia: A Sioux' head, as 20 Smaldeel, but circled in green.
Motto: Tenacité (Tenacity)
Status: Not currently formed

With the arrival of the C-119G Boxcar for 15 Wing a third squadron was formed to share the type with 20 Smaldeel, and this was established on 1 April 1954 as 40 Transportsmaldeel, based at Melsbroek. Following the pattern set by the other two squadrons of 15 Wing, a Sioux' head was adopted as the unit insignia, but in this instance, circled in green.

40 Smaldeel remained in commission only until June 1954 when it was disbanded as an economy measure and the Boxcars re-grouped under 20 Smaldeel. Between 1957 and 1960, the green Sioux was used by the Transport Flight at Kamina, Congo, but 40 Smaldeel was re-formed in June 1961 and once again claimed the insignia. With the arrival of the Hercules, the Boxcars were placed in storage and 40 Smaldeel disbanded on 31 June 1973. The unit was later made a helicopter Smaldeel under a different insignia.

40 HELI-SMALDEEL

Insignia: A white Sea King helicopter in a blue disc above a red hand, surrounded by a white circle with black lettering, golden crown and palm leaves.
Motto: Aude Audenda
(Dare the things that need to be dared)
Status: Currently operate Sea Kings, HSS-1 and Alouette III from Koksijde and elsewhere on detachment.

Originally formed at Koksijde as the Search And Rescue Flight on 1 April 1961 the unit replaced high-speed motor launches used for North Sea Rescue. Following crew training by the French Navy unit 23S at St. Mandrier, five Sud-built HSS-1 were received, being later joined by seven ex-SABENA S-58Cs.

In 1963, a second flight (the SRT Flight) was established, with additional responsibility for teaching helicopter transport techniques – the two flights forming the 'Helikoptersmaldeel'. A further two sections were added later, being the Zeemacht Flight attached to the Belgian Navy, and the Onderhouds Flight for support and maintenance of the resident helicopters.

In 1971 three Alouette III re-equipped the Zeemacht Flight, and on 30 October 1974 the designation 40 Smaldeel, late of 15 (Transport) Wing was applied to the four sections. During 1976 the SAR Flight received its Westland-built Sea Kings although the SRT Flight continues to provide support facilities for the 9 and 13th Missile Wings in Germany with the ageing S-58.

41 SMALDEEL

Status: Not formed

42 SMALDEEL

Insignia: A red, winged devil, looking through a telescope.

Motto: Nil
Status: Currently equipped with Mirage 5BR aircraft, based at Florennes. Part of 2 Wing.

Aerial photography of an experimental nature was practised for the first time on 23 September 1914 by Pilot Deschamps and Observer Schmidt, and this soon led to a 'Foto Sectie' being established at Evère in 1919 and this later became 7 Verkenningssmaldeel, IV Groep, I Regiment (7VSm, IV Gr, 1 Rgt.) at Goetsenhoven, equipped with Foxes, at the outbreak of World War 2.

At Wahn, the 42 Tactische Verkenningssmaldeel was formed in June 1953 as 'C' Flight of 2 Squadron, RAF. They flew camera-equipped F-84G Thunderjets (with 'H8- ' codes) alongside the RAF Meteor FR.9s. As Belgium's sole reconnaissance squadron, it naturally adopted the 1 Verkenningssmaldeel (1 VSm.) insignia.

The Flight became 42 Smaldeel in 1955 and in the August it converted to RF-84F Thunderflashes. In June 1957 the squadron moved to Brustem, but it was to lose its autonomy when moved to Bevekom as part of 1 Wing, in November 1960. On 1 April 1963 another move took the squadron to Bierset, where it became part of 3 Wing from 1968 – remaining there until 15 September 1971, when partially equipped with the Mirage 5BR, it joined 2 Wing at Florennes and completed conversion.

The squadron left its Thunderflashes to 1 Smaldeel at Bierset, where they continued to provide TacRecce coverage until 42 Smaldeel became operational, staging a final formation flypast on 4 May 1972.

43 – 49 SMALDEEL

Status: Not formed

50 SMALDEEL

Insignia: A blue and gold swallow on a red circle
Motto: Nil
Status: Currently a component of 13 Wing Missiles, equipped with Nike-Hercules missiles and based at Düren.

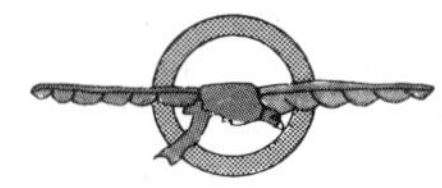

The swallow insignia originated in 1918 on the aircraft of 5 Verkenningssmaldeel at Bray Dunes. Following the First World War, 5 Verkenningssmaldeel established itself at Goetsenhoven where in 1930 it became the first squadron to re-equip with the Fairey Fox. 5 Smaldeel was still flying Foxes in 1940 as the sole squadron of III Groep, 1 Regiment, when the German invasion forced disbandment.

In December 1953, 25 Dagjachtsmaldeel was formed with 'VT- ' coded Meteor F.8s in 13 Wing at Koksijde, continuing to fly these aircraft at Brustem until disbanded on 1 July 1958.

When 13 Wing re-mustered as a missile unit, its 50 Smaldeel, which formed in May 1959, adopted the swallow in the form previously used by 5 and 25 Smaldelen.

51 and 54 SMALDELEN

Insignia: An encircled eagle
Motto: Quarens Quem de Voret
Status: Currently component units of 13 and 9 Wings, equipped with Nike-Hercules missiles, one being based at Blankenheim.

In 1912 the Compagnie van Luchtschippers en Mecaniciens (Airship Crew and Support Company) was split into two sections; the Luchtvaart (Airservice) and Luchtscheepvaart (Airship service) the latter having control over fixed (fortress) and dirigible balloons and airships. Three dirigibles, the Zodiac, Belgie and Stad Brussel were joined in 1912 by two Parsevals, but from 1914 until disbandment in 1934, French types, notably the Caiquot, were used for battlefield surveillance.

To identify the airship service, the eagle badge was authorised on 1 May 1917 in three forms, gold for officers, silver for warrant officers and sergeants and bronze for corporals and men.

In 1935 the establishment of two squadrons at Nijvel to form III Groep of 2 presented the Regiment opportunity to perpetuate the famous eagle badge, and 5 Smaldeel employed a blue 'O' and 6 Smaldeel , a red. Both units were equipped with Foxes, but awaiting the Brewster 339 (Buffalo) at the time of the German invasion.

When 1 Wing required a third Meteor F.4 Squadron, 4 Smaldeel was established at Bevekom in December 1949 with 'SV- ' codes and the red-circled eagle, later progressing to the Meteor F.8 in 1951, but disbanding in September 1957 after its sister units had received Hunters. The unit was revived in May 1959 as 51 Smaldeel, 13 Wing equipped with Nike missiles.

The blue-circled eagle was adopted by 22 Smaldeel of 9 Wing formed at Bierset on 2 September 1953, flying 'IS- ' coded Thunderjets, Meteors and later the Hunter F.6 before disbanding on 30 April 1960. Its insignia was likewise allocated to a Nike squadron, in this instance 54 Smaldeel of 9 Wing on 28 May 1962.

52 SMALDEEL

Insignia: A holly leaf within a bronze coloured belt
Motto: Qui s'y Frotte s'y Pique
(Who provokes it, is hurt)
Status: Currently a component of 13 Wing, equipped with Nike-Hercules missiles and based at Euskirchen.

First seen on the Breguet 14 aircraft of 3 Verkenningssmaldeel in 1918, the holly-leaf insignia was perpetuated between the wars, and appeared on the Fairey Foxes of 3 Smaldeel, II Groep, 1 LvRegt., at Goetsenhoven in 1936-40, with the old motto 'N'Importe Ou'.

In January 1950, the emblem returned on the Spitfires of 3 Jagers-Bommenwerperssmaldeel, 2 Wing at Florennes, later with 'YL- ' coded Thunderjets and in August 1955 the first of the Belgian Thunderstreaks. The squadron was disbanded on 15 October 1960 and the badge transferred to 52 Smaldeel of 13 Missile Wing which had formed with Nikes in May 1959.

53 SMALDEEL

Insignia: A golden dragon
Motto: Nil
Status: Currently a component of 9 Wing, equipped with Nike-Hercules missiles and based at Kaster, in West Germany.

Taken from the personal insignia of a Farman pilot, the dragon was adopted in 1918 by 7 Verkenningssmaldeel at Houtem, but with the reorganisation of the air force in 1926, 1 Verkenningssmaldeel employed the emblem in gold and 3 Verkenningssmaldeel, in silver. Both squadrons comprised 1 Group of 3 Regiment and in 1940 were equipped with the Fairey Fox at Evère.

With the arrival of the Mosquito NF.30 at Bevekom, the Nachtjacht Flight formed on 25 May 1948 and was expanded to form 10 Smaldeel in 1950 with aircraft coded 'KT- ', in 1 Jachtwing Alle Weder (All-weather Fighter-Wing).

10 Smaldeel employed the silver dragon insignia and marked this on its later Meteor NF.11s until disbanded on 15 September 1957. The gold dragon was brought into use once more when 53 Smaldeel, 13 Wing formed in May 1959 at Lombardsijde equipped with Nike missiles. The squadron moved to Kaster in West Germany following its working-up period and transferred to 9 Wing on 12 August 1964.

54 SMALDEEL

See 51 Smaldeel

55 SMALDEEL

Insignia: A red arrow with silver wings
Motto: Nil
Status: Currently a component of 9 Wing, equipped with Nike-Hercules missiles and based at Kapellen-Erft.

Between 1937 and 1939, fighter-reconnaissance squadron, 7 Jachtverkenningssmaldeel, II Groep, 3 Regiment applied a 'Cupid' insignia to their aircraft with the motto 'Parcere Subjectis et Debeliare Superbos'.

A simplified arrow motif was adopted in 1939, and at the outbreak of war the unit was at Evère under III Groep of 3 Regiment flying the Fairey Fox and awaiting deliveries of SABCA-built Breguet 694 twin engined monoplanes.

It was not until December 1953 that the arrow re-appeared as a marking on the 'KS- ' coded Meteor F.8 aircraft of 33 Dagjachtsmaldeel, 13 Jachtwing at Koksijde (later, Brustem). A reorganisation in 1957 brought about the dissolution of the third squadron in each wing, with the result that 33 Smaldeel disbanded on 2 May 1957.

Five years later, on 28 May 1962, 55 Smaldeel of 9 Wing was formed with the Nike missile inheriting the arrow and its historical connection with 7 Smaldeel.

56 SMALDEEL

Insignia: A silver cygnet made of folded paper in a silver circle.
Motto: Nil
Status: Currently a component of 9 Wing, equipped with Nike-Hercules missiles and based at Grefrath.

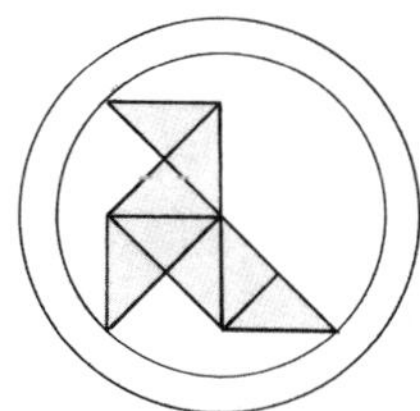

One of the most famous markings in the air force, the cygnet was the personal emblem of air ace Willy Coppens and appeared on his Nieuport Bébé. It was later applied to the aircraft of 9 Jachtsmaldeel in 1917, but when 9 Smaldeel adopted the thistle motif the cygnet was taken over by 11 Jachtsmaldeel.

11 Smaldeel returned to Schaffen in 1923 after being a part of the occupation forces in Germany and was divided into two parts; 3 Jachtsmaldeel having a red cygnet circled in gold, and 4 Jachtsmaldeel with a silver cygnet circled in silver. On the outbreak of war, both units comprised II Groep of 2 Regiment and flew the Fiat CR-42, a recent replacement for the Fairey Firefly.

The silver emblem of 4 Jachtsmaldeel was put into use once more on 15 November 1963 when it was adopted by the Nike missile unit 56 Smaldeel of 9 Wing.

57 SMALDEEL

Insignia: A black and white zebra on a globe
Motto: Nil
Status: Currently a component of 13 Wing, equipped with Nike-Hercules missiles.

One of the oldest insignias of the BLu – the zebra – was used in 1918 by the Spads of 4 Verkenningssmaldeel at Hondschoote. Following re-organisation of the Aviation Militaire 4 Verkenningssmaldeel was established at Goetsenhoven in II Groep of I Groepering in 1923 but disbanded shortly afterwards.

In 1977 the 4 Verkenningssmaldeel badge was adopted by 57 Smaldeel of 13 Wing as the eighth Nike-Hercules squadron.

349 SMALDEEL

Insignia: Two 'morning stars' crossed, on a white background
Motto: Strike Hard, Strike Home
Status: Currently a component of 1 Wing, flying TF- and F-104G Starfighters at Bevekom.

349 Squadron, RAF, formed at Lagos on 10 November 1942 with Belgian personnel intending to fly Tomahawks in the Congo. It was instead transferred to Collyweston on 5 June 1943 and re-equipped with Spitfires coded 'GE- ', later participating in the liberation of Europe during 1944-45.

When the war ended in May 1945, 349 and 350 Squadrons were at Wunsdorf operating as 123 Wing. Both transferred to Fassberg under the designation 160 Wing, and finally returned to home ground on 24 October 1946 at Bevekom to become part of the Belgian air force.

On 1 February its parent unit was re-numbered 1 Wing and in the next year, 349 Smaldeel followed 350 in adopting the Meteor F.4. The Meteor F.8 arrived in 1951, Hunters in March 1957, and CF-100 Canucks only nine months later. These were retained until the arrival of Starfighters in the Summer of 1963.

350 SMALDEEL

Insignia: The head of Ambiorix in blue and gold on a white circle.
Motto: Belgae Gallorum Fortissmi (Belgians, bravest of the Gauls)
Status: Currently a component of 1 Wing, flying TF- and F-104G Starfighters at Bevekom.

Formed as the first of the two Belgian squadrons of the RAF, 350 Squadron came into being at Valley on 12 November 1941 flying Spitfires, and later took a part in the liberation of Europe, returning to Belgian soil at Evère on 4 December

1944. Much of the squadron's equipment came from patriotic individuals in the Congo who collected £250,000 to buy 50 Spitfires.

350 Squadron saw the cessation of hostilities together with 349 Squadron at Wunsdorf and under the joint title of 123 Wing. After a transfer to Fassberg as 160 Wing, the two squadrons left the RAF on 15 October 1946, and returned to

Bevekom, Belgium, on the 24th, where 160 Wing was re-numbered 1 Wing on 1 February 1948. Meteors replaced the Spitfires in June 1949, the F.4 giving way to the F.8 in July 1951, and these were followed by Hunters in March 1957, Canucks in May 1958 and Starfighters. Following the departure of the Canucks in August 1962, 350 remained without aircraft until it became the first Belgian Starfighter squadron when its initial aircraft arrived in April 1963. The type remains in use as an air-superiority fighter and will be replaced by the F-16.

351 SQUADRON

Insignia: A green thistle surrounded by a belt
Motto: Nemo Me Impune Lacessit
(No man provokes me with impunity)
Status: Not currently formed

The first operational squadron to be formed in Belgium after World War 2, adopted a number immediately following the two Belgian units of the RAF, 349 and 350 Squadrons, although this had previously been used by a Yugoslav/RAF squadron. 351 Squadron formed at Florennes on 1

November 1946 as a fighter unit, equipped with the Spitfire XIV, and adopted the code '3R- ', and thistle insignia of 1 Smaldeel. On 10 January 1948, the Squadron took up its old designation as 1 Jagers-Bommenwerperssmaldeel and its parent unit became 2 Wing on 1 February 1948.

352 SQUADRON

Insignia: A red comet of five points and tail
Motto: Ut Fulgur Sulca Aethera
(Cleave the air like lightning)
Status: Not currently formed

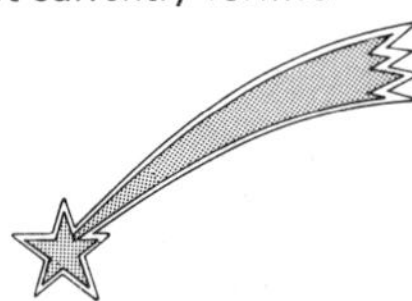

352 Squadron joined 351 Squadron in 161 Wing at Florennes early in 1947, marking its aircraft with 'VR- ' codes and the 'Comet' of the World War 1, 5 Jachtsmaldeel. The squadron number duplicated that of a Yugoslav unit in the RAF, but had no connections with its forbear. On 1 November 1947, 352 Squadron and its Spitfire XIV equipment was re-numbered 2 Jagers-Bommenwerperssmaldeel and its Wing was designated 2 Wing on 1 February 1948.

353 – 365 SQUADRONS

Status: Not formed

366 SQUADRON

Insignia: Nil
Motto: Nil
Status: Not currently formed

366 Squadron formed at Evère with Dakotas on 15 August 1946, from the Belgian-manned 525 Squadron (RAF), and together with 367 Squadron comprised 169 Wing. It was re-numbered 20 Smaldeel on 1 February 1948 and received the well-known Sioux' head emblem thereafter.

367 SQUADRON

Insignia: Nil
Motto: Nil
Status: Not currently formed

Formed as one of 169 Wing's two squadrons in August 1946, 367 Squadron operated as a light communications unit and a successor to the Belgian section of the Allied Flight at Hendon, equipped with a mixture of Ansons, Dominies, Oxfords, Proctors, Hurricanes and a Magister, at Evère. The squadron became 21 Smaldeel on 1 February 1948.

368 SQUADRON

Status: Not formed

369 SQUADRON

Insignia: Nil
Motto: Nil
Status: Not currently formed

The first army co-operation squadron of the post-war Luchtmacht formed on 31 July 1947 to receive Austers delivered new from Britain. Pending the completion of repairs at Brasschaat, the Austers were delivered to store at Wevelgem and not put into use until 1948.

On 1 February 1948 the RAF-style numbering of the Belgian units was discontinued and 369 Squadron became 15 Smaldeel.

BASIS VAN DE LUCHTMACHT

Insignia: In a shield, a propeller, a half cogwheel and two wings
Motto: Nil
Status: Currently controls the 21, 22, 23 and 25 Wings, the Mekanografisch Centrum and the Steunwing Evère, from Gent.

The Basis van de Luchtmacht is one of the three principal air force commands and since 1961 has been situated at Gent. It contains units with a principally logistic or administrative function. The first group contains the Logistic Wings and their stocks of supplies, provisions and replacements. In the second group is the administration wing, which until June 1976 was stationed at Gervizet Barracks, Etterbeck and entrusted with the administration of personnel in the service of the General Staff at home and abroad. At Evère is the General Service Centre of the Basis Luchtmacht, controlling accommodation and catering facilities, whilst the General Service Unit at Gent administers the Basis Luchtmacht headquarters.

The functions of the second administrative group were taken over in June 1976 by the Steunwing Evère (Evère Support Wing).

The Basis Luchtmacht controls the 21, 22, 23, 25 Wings, the Mekanografisch Centrum and the Steunwing Evère.

BELGISCH PERMANENT DETACHMENT, SOLENZARA

Insignia: Three red aircraft, signifying one each from Belgium, France and Germany, and a blue and silver NATO emblem; all superimposed on a map of Corsica.
Motto: Nil
Status: Current

Armament practise is regularly carried out at Solenzara by detachments of aircraft from the Luchtmacht, Luftwaffe and Armée de l'Air.

The above insignia is that of the ground section responsible for administration and servicing of Belgian detachments.

THE 'BLUE BEES'

Insignia:
Motto:
Status: Currently a helicopter display team.

For a number of years the team was equipped with Piper Cubs, but in 1968 a new machine, the Alouette II was adopted and at the same time the title 'Blue Bees' was assumed. The team pilots and servicing crew are all volunteers from the 17th Light Aviation Squadron, stationed at Werl in West Germany.

CENTRUM VAN MILITAIRE VORMING

Insignia: A black Starfighter forming a blue 'C' in which is a blue torch with yellow flame and a brown and black rifle; the whole superimposed upon a white shield with silver borders.
Motto: Nil
Status: Currently trains recruits in airfield defence procedures, at Koksijde.

In February 1946 the 'Receiving Centre' for material used in forming the new air force was moved from Snailwell and Bottisham to Bevingen, and later in the year to St. Niklaas-Wass where in 1948 it was renamed Instructie Centrum van de Luchtmacht. A move to Melsbroek also in 1948 brought about a change of name to Recruterings-centrum, Instructie en Depot, and between 1950 and 1956 when the unit was at Evère it was known as the Instructiecentrum voor Vliegveldverdedigingseenheden (Training Centre for Airfield Defence Units).

From 1957 the centre was split into two; 1 Instructiecentrum at Flawinne for French speaking recruits, and 3 Instructiecentrum at Turnhout for Flemish speakers, these two coming under the control of the Technische School van de Luchtmacht at Saffraanberg in 1960. Both were com-

bined at Chièvres on 2 February 1964 as the Centrum voor Militaire Vorming (Militaire Induction Centre) and on 17 May 1967 the unit moved to Koksijde where its principal task is the training of recruits in airfield defence procedures.

DIENST CONTROLE EN TECHNISCHE KEURING

Insignia: In a blue cogwheel on a yellow disc, a blue aerodynamic shape with red airflow lines. Also the initials of the unit's French designation 'Service de Controle et de Réception Technique'.
Motto: Nil
Status: Currently an aircraft acceptance unit.

Prior to 1940, the air force saw the requirement for an aircraft acceptance unit having as its principal task, 'quality control'. This was established after World War 2 as the DCTK, its staff being attached to firms having air force contracts, although it additionally has a laboratory at Evère.

The initials of the units French designation Service de Controle et de Réception Technique also appear on the insignia.

DIENST VOOR PROEFVLUCHTEN

Insignia: On a yellow disc, a blue cogwheel and a blue aerodynamic shape with red airflow lines.
Motto: Nil
Status: The 'test-flying' service, currently based at Gosselies.

The 'Test-flying Service' of the Luchtmacht was established at Wevelgem in 1946 by Major Legrand and Commandant Eyskens both of whom had received training at the RAF's Empire Test Pilots School. Four pilots are currently attached to the DPV and are principally involved in test-flying aircraft built or overhauled by SABCA and Avions Fairey.

ELEKTRON SMALDEEL - 10JBW

Insignia: A blue disc, containing a red lion and golden electric beams on a yellow inner circle. Black lettering with gold and red crown.
Motto: Audax et Cautus (Bold yet cautious)
Status: Currently based at Kleine Brogel. No aircraft allocated.

The Elektron Smaldeel is responsible for all radar, communications and other electrical equipment operated by 10 Wing. Similar units are attached to the other Luchtmacht Wings.

ELEMENTAIRE VLIEG SCHOOL

Insignia: A black and white penguin with red eye and yellow beak and feet.
Motto: Nil
Status: Currently equipped with SF-260M at Goetsenhoven

On 9 September 1910 the War Ministry issued its first instruction to the new Militaire Luchtvaart, commanded by Le Clément de Saint Marcq, ordering Leutenant Georges Nelis to Kiewit to commence basic pilot training under Chevalier de Laminne, and later to Mourmelon for advanced instruction.

In 1911 the training centre moved to Brasschaat but on 29 March 1913 it was decided to concentrate all training at Kiewit where it remained until transferred to Wilrijk near Berchem. The German advance forced evacuation to Beaumarais near Calais in October 1914.

Following the First World War the school moved to As and later Wevelgem, before training schools were set up at Deurne, Gosselies and Saint Hubert. From 1931, Belgium sought after a more complete training programme, but it was not until 1939 that the Vliegschool was established comprising Schoolsmaldelen at Deurne, Gosselies, Saint Denis Westrem and Wevelgem.

Hardly had this step been taken than the training units were forced to evacuate Belgium on 12 May 1940 and re-group at Oujda, Morocco on 1 June, from where many pilots made their way to England.

In preparation for the re-establishment of the Belgian Air Force, the Elementaire Vliegschool,

formed at Snailwell in April 1945 with thirty RAF Tiger Moths, taking these home to Schaffen on 9 March 1946. Twenty home-produced Stampe SV-4B augmented the Tiger Moths during 1948 and on 22 July 1950 the school moved to its present base at Goetsenhoven. Here, forty-five more Stampes were received between 1951 and 1955 remaining in service until replaced by the SIAI SF-260M in 1971.

FORCE AERIENNE TACTIQUE CONGOLAISE

Insignia: In a blue disc, a brown barrow with black wheels, holding a brown packing-case and barrel, a white and red first-aid box and a brown sack. The barrow is propelled by a pair of wings.

Motto: Nil

Status: Not currently formed

In 1964, the Congo Prime Minister Mr Tschombe, asked the Belgian government for logistical supplies to support the troops now maintaining order after the civil war. Approval was given for 15 Wing to commence regular supply flights to the old Kamina airfield which was specially re-opened. Additionally, the Belgians supplied crews for the DC-3 and DC-4 aircraft of the Congolese Air Force until the assistance programme was completed in 1967.

GROEP VLIEGVELD BEVEILIGING - 10JBW

Insignia: A brown fox head on a green background the figure '10' on a blue background, two green wings and the Royal Crown in red, white and green. Lettering in black on a green ribbon.

Motto: Nil

Status: Currently based at Kleine Brogel. Has no aircraft on charge.

The GVB operates in the same manner as the RAF Regiment in providing defence for Luchtmacht air-

fields, in addition to everyday security patrols. Similar units are attached to most other wings.

GROEPERING MISSILES

Insignia: A White Nike-Hercules missile with red flames and a black star, all on a shield divided into dark blue and light blue sections.

Motto: Nil

Status: Currently administers the missile wings of the Belgian Air Force. Based at Grefrath, West Germany.

The Groepering Missiles is the administrative unit for missile wings of the Belgian Air Force and comprises three wings, 9 Wing, 13 Wing and the Wing Support Missiles at Düren. Two S-58/HSS-1 from the SRT Flight of 40 Smaldeel are permanently attached to the missile wings.

GROEPERING OPLEIDING EN TRAINING

Insignia: A dark blue torch with yellow flame and red edges, also a pair of white wings on a blue shield with yellow edges.
Motto: Nil
Status: Current, based at Evère.

The GOT administers all training establishments in the air force, having control over the Vervolmakings Centrum, the Centrum Militaire Vorming and the Technische School.

HULP SMALDEEL

Insignia: A black hand holding a yellow torch with red flames and flanked by two orange wings; the whole on a white disc, which is itself surrounded by a blue circle with golden edging.

Motto: Nunquam Flamma Extinguitur (The flame is never extinguished)
Status: Not currently formed.

The 'Help Squadron' was established at Bevekom on 1 December 1949 under 1 Wing equipped with the Spitfire IX and XIV (codes 'GV- ') and staffed by civilian weekend pilots as an auxiliary squadron. Following the arrival of Meteor F.4s, the squadron moved to Brustem on 1 April 1957 coming under the administrative control of 13 Wing. It was disbanded on 1 February 1958.

JACHT SCHOOL

Insignia: A yellow and gold falcon holding a red and black arrow and encircled in blue.
Motto: Nil
Status: Not currently formed

The Jacht (Fighter) School was established at Koksijde on 12 January 1948 equipped with 'IQ- ' coded Spitfires and Martinet target tugs, to provide continuation training to pilots completing their Harvard and Oxford training at St. Truiden.

In 1953 the unit became the Jachtvliegschool (Fighter Pilots' School) and divided its new equipment into two flights: the Reconvertiesmaldeel with the Meteor T.7 for twin-seat conversion, and the Operationele Eenheid (Operational Unit) with Meteor F.4s acting as an OCU. The school later received the T-33A and Meteor F.8, but disbanded on 15 July 1958 when a joint Belgian/Dutch training programme was initiated at Brustem.

LUCHTMACHT OFFICIEREN SCHOOL

Insignia: A parrot with white plumage and yellow head, black beak and red body, wearing black spectacles and holding a yellow scr scroll.
Motto: Nil
Status: Not currently formed.

A Luchtvaartschool (Aviation School) was established at Evère in 1930 and the next year the parrot insignia was chosen by Baron Thierry d'Huart. The principal task of the school was the training of officer cadets, and a separate school, with the same insignia, was opened at Zellik to train NCOs.

After closure in 1940, the officer school reopened on 18 November 1946 at Nijvel, later combining with the Luchtopleidings Wing to form the Vervolmakings Centrum van de Luchtmacht (Air Force Advanced Training Centre). The school was finally closed on 27 March 1962 and instructional responsibility passed on to the Instructie Groep/Vervolmakings Centrum.

MAINTENANCE GROEP - 10JBW

Insignia: A red and gold Atlas, holding a white and gold globe on a blue background, over which are three gold birds. Gold lettering on black.

Motto: Oculus ad Omnia (An eye on everything)
Status: Currently based at Kleine Brogel, though has no aircraft on charge.

The 10 JBW Maintenance Group was formed on 29 November 1951 and adopted the Atlas insignia in 1962. The group provides maintenance facilities for the squadrons attached to 10 Wing.

MEKANOGRAFISCH CENTRUM

The CEMEC is a new and specialised department of the Basis Luchtmacht which with the aid of a computer investigates methods of improved stock control, accountancy and the like in the interests of economy and efficiency. CEMEC was formed at Evère in 1975.

METEOROLOGISCHE WING

Insignia: A white cloud over a yellow rising sun.
Motto: Nil
Status: Current

A meteorological section was formed for the Belgian squadrons of the RAF in 1942 and disbanded in 1946. However, on 1 January 1947 the Meteorologische Dienst was re-established and in 1949 made responsible for providing forecasts to all three services. A rise in status to Korps level occurred on 2 December 1954 and on 15 October 1966 the present title was allocated.

NAVIGATIE SCHOOL

Insignia: A white seagull above a blue sea and within a red circle.
Motto: Au Large
Status: Not currently formed as navigator training is now undertaken in the U.S.A.

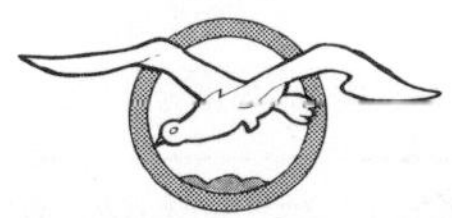

The seagull insignia of 2 Verkenningssmaldeel originally adopted in 1917 at Les Moeres, reappeared at Goetsenhoven in 1926 with 1 Verkenningssmaldeel, II Groep, I Regiment, remaining with 1 Smaldeel when it was transferred to I Groep, I Regiment and re-equipped with Foxes in 1936. In May 1940 the squadron was destroyed during the German invasion whilst flying from Deurne.

The Navigatieschool established in May 1948 revived the seagull, and until disbanded on 1 April 1955 was responsible for the training of air force navigators in Pembroke aircraft loaned from 15 Wing and based at Melsbroek. Navigator training is currently undertaken in the USA.

In 1964, plans were made to establish a helicopter transport flight under the old 1 Smaldeel insignia, but in the event, these were not proceeded with.

NACHTJACHT FLIGHT

See 53 Smaldeel

PROGRAMMING & TRAINING CENTRE

Insignia: On a green and ivory chessboard, a castle surrounded by a red ring and marked in black with initials 'PTC'. Motto on a blue ribbon.
Motto: Sic et Non (There and yet not there)
Status: Not currently formed

The PTC was established at Glaaien in 1966 as one of the four programming units in NATO.

Until disbandment, the centre instructed trainee radar operators.

RIJKSWACHT

Status: Currently equipped with Alouette Astazou and SA.330 Pumas.

The Rijkswacht is a unit combining the duties of civil defence, police and coastguard based alongside SLV/15 Smaldeel at Brasschaat, its title meaning literally 'National Guard', and its function being equivalent of the Gendarmerie. For reasons of economy Rijkswacht helicopters are maintained by the army, the first of six Alouette Astazous having arrived in 1968. These were followed in 1973 by three SA-330 Puma, although five were reportedly ordered originally.

First use of aircraft by the Rijkswacht was in 1950 when Auster AOP.6 A-14 was employed on traffic control work in the Brussels area. Other aircraft were later borrowed from the Landmacht as required but it was not until the arrival of the Alouettes that the unit officially possessed its own equipment.

RODE DUIVELS

Insignia: A red devil mask and four red Magisters all on a blue disc.
Motto: Nil
Status: Currently employ specially decorated Magister aircraft of No's 7 & 9 Smaldeel.

Prior to the Second World War, Belgium had several aerobatic teams flying Avro, Fiat, Firefly, Fox, Gladiator and Morane types. The aerobatic tradition was perpetuated in 1951 when 1 Wing at Bevekom formed the 'Acrobobs' with their newly received Meteor F.8s under Captain Bobby Bladt. A year later the team won an aerobatic competition at Lyon beating entries from several other nations.

In 1957 Major Bladt was at Chièvres where five of the new Hunter F.6 aircraft of 7 Wing formed the Rode Duivels/Diables Rouge (Red Devils). The number of all-red Hunters increased to sixteen in 1959 but the ever-present danger of a multiple collision reduced this in the next year to four plus one solo performer. A final performance on 4 October 1963 preceeded the dissolution of the team and disbandment of 7 Wing.

During 1965, six red-painted Magisters of the VVS took over the vacant aerobatic team position, led by Major De Waelheyns and resurrected the Red Devils title. The first display was given on 27 June and shortly afterwards the team joined the Patrouille de France – also with Magisters – for a combined performance. A reserve pilot was provided by the Dutch Air Force, and in return, Belgium supplied an extra crew member for the 'Whisky Four' team of Thunderstreaks.

The Rode Duivels aircraft are used by 7 and 9 Smaldelen when not required for aerobatic work and are decorated with a lion insignia either side of the front cockpit.

SEMMERZAKE RADAR OPERATING SYSTEM and OPSPORINGS EN CONTROLECENTRUM GLAAIEN

Semmerzake insignia: A dark blue arrow, two dark blue wings and two red lightning flashes all on a black shield with red borders.
Glaaien insignia: As above but with light blue arrow and yellow wings.
Status: Current - see below

In 1948 a small group of technical specialists formed the first experimental radar section in the Luchtmacht at Brustem, and in April 1949 installed it in the famous Eben Emael fort. Here, it was used to train a cadre of radar controllers for the 1 Ground Control Interception Unit established at Bevekom in February 1950, when British GCI units were able to hand over responsibility to the Belgians.

1 GCI moved to Glaaien on 19 July 1951 and in 1955 was housed in an underground bunker there, where it became 2 Opsporings en Controlecentrum (2 Reporting and Control Centre). The unit took responsibility for GCI in the Belgian sector until December 1961, when the German-Belgian sector of the middle European area passed to the control post at Uedem in Germany, whereupon 2 OCS became OCS Glaaien.

In addition to the Opsporings en Controlecentrum, is the Opsporings en Controle Post at Semmerzake, linked to the NATO early-warning system NADGE. The first Belgian early-warning radars took over from RAF mobile units in 1950 at Maldegem, moving shortly afterwards to Koksijde and later Semmerzake where underground quarters were built in 1956.

Sharing the Semmerzake base with the Opsporings en Controle Post was the Traffic Co-ordination Centre (TCC), responsible for the integration of civil and military air traffic. This moved from Glaaien on 22 March 1966 and a year later introduced its first civilian controllers and civil information service. Jointly known as the SEROS (Semmerzake Radar Operating System) the Traffic Co-ordination Centre and Opsporings en Controle Post co-ordinate their actions although the former is purely national, and the latter a NATO unit under the Tactische Luchtmacht.

Until 1973, Semmerzake was the home of the School van Opsporings en Controle, the unit formed at Eben Emael in 1949. In 1954 this control and reporting school for radar operators moved to Saffraanberg and in 1959 to Semmerzake where practical instruction could be given to students. The school disbanded on 21 December 1972 and on 1 January 1973 was re-established at Erndtebruck, the international traffic control centre for Holland, Belgium and Northern Germany.

SCHOOL VAN HET LICHT VLIEGWEZEN

See 15 Smaldeel

SLEEPVLUCHT

See 'Target Towing Flight'

SLIVERS

Insignia: In a blue oval, the badge of 1 Wing, the Belgian roundel, two grey Starfighters producing a red-yellow flame and in black lettering 'SLIVERS BELGIUM'
Motto: Nil
Status: Not currently formed

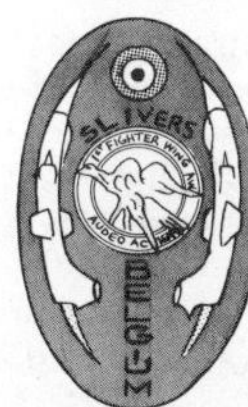

The Slivers team of two Starfighters formed in 1 Wing in 1968 and gave their first public display at Bevekom on 14 May 1969. The duo were the official F-104G display team and regularly appeared at displays throughout Europe until disbanded in 1975.

SMALDEEL VLIEGEN ZONDER ZICHT

Insignia: A white marabou in an orange circle, surmounted by a red and gold crown, green palm leaves. Lettering in black.
Motto: Non Sibi (Not for themselves)
Status: Currently equipped with T-33As, based at Brustem.

The badge of a white Marabou in a red triangle was first applied to aircraft of 2 Verkenningssmaldeel, IV Groep, I Luchtvaartregiment at Evère in 1930 and later adopted by 6 Smaldeel, III Groep, III Regiment.

An Instrument Flying School, the smaldeel Vliegen Zonder Zicht, was formed in 1958 as a section of the Vervolmakings Centrum and equipped with the ex-Jachtschool T-33As. It adopted the Marabou insignia at Bierset in November 1966, and in 1968 was combined with 42 Smaldeel at the same base to form 3 Wing. The Vliegen Zonder Zicht transferred to Brustem on 14 May 1971 and the next month was re-designated 11 Smaldeel.

STEUNWING EVERE

Status: Current - see below

Newest of the six sections of the Basis Luchtmacht, the Steunwing Evère was formed at Evère in June 1976 to combine the administrative groups of the air force, concerned particularly with personnel. The Steunwing (Support Wing) thus controls accomodation and catering units and provides staff for the administrative departments.

SWALLOWS

Insignia: Two animated aircraft within a circle
Motto: Nil
Status: Currently fly three EVS-owned SF-260M as an aerobatic display team from their base at Goetsenhoven.

With the arrival of the Siai-Marchetti SF-260M in 1970 several pilots began practising aerobatics and in 1972 Adjutant Serrien and Onder-Lieutenant Delacuvellerie performed at several displays, forming the Swallows team the next year, with Serrien and Kapitein Fonderi. In 1975 the team again changed with the return of Delacuvellerie and the addition of Adjutant Goussens.

The Swallows are the successors to the 'Pinguins' team of two Stampe SV-4 biplanes, renowned during the sixties for their 'mirror flying'.

TARGET FLIGHT - 10 JBW

Insignia: A white bird with yellow wings and orange beak, also a red devil with black hair, white trousers and horns and holding a white Starfighter in the hand, and an orange tiger.

Motto: Nil

Status: Currently based at Kleine Brogel, flying the F-104G Starfighter

10 Wing formed a target flight in January 1969, which operates gunnery and target practise aircraft for exercises at home and on detachment to Sardinia.

TARGET TOWING FLIGHT (SLEEPVLUCHT)

Insignia: On a blue shield with black borders, an orange and blue lighthouse standing on two brown rocks in a green sea, and projecting a yellow light beam.

Motto: Nil

Status: Not currently formed

Commencing in 1947, target towing was undertaken by Miles Martinets based at Koksijde, but in 1948 the establishment of the Jacht School (Fighter School) here brought Mosquitos to further the work.

In March 1954, the Target Towing Flight was formed at Sylt, from 24 Smaldeel (codes 'B2- ') with crews and technical personnel from the squadrons which visited for gunnery practise. The name Luchtmacht Detachment van Sylt was adopted on 31 May 1955. On 1 June 1957 the 'Flight Target Towing' at Sylt had twelve Meteor F.8s and was placed under the administrative control of 13 Wing at Brustem. The unit disbanded on 30 June 1960, and on 22 August the Flight was re-established at Decimomannu, Corsica with eight Meteor F.8 aircraft and a few Canucks, using the same home-base at Koksijde. The Flight disbanded on 9 November 1963, since which time detachments to Corsica have provided their own drogue aircraft.

TECHNISCHE SCHOOL (KAMINA)

Insignia: A propeller, a black and gold cogwheel and the golden star of the Congo, all on a white disc encircled in red, with gold edging.

Motto: Labor Omina Vincit Improbus (Persistant work overcomes all things)

Status: Not currently formed

The Kamina-based Technical School was divided into two sections: the first having the task of imparting a basic technical knowledge to recruits and the second giving specialist training in the fields of engines, instruments and electronics. As a result of the civil war, the school closed in 1960.

TECHNISCHE SCHOOL VAN DE LUCHTMACHT

Insignia: A propeller over a red cogwheel, flanked by white wings all in a yellow triangle with black edging, which is itself superimposed on a blue shield, bordered in yellow.

Motto: Labor Omina Vincit Improbus (Persistant work overcomes all things)

Status: Currently has several non-flying aircraft on charge, at its base at Saffraanberg.

The air force technical school was originally established at Snailwell, England in 1944 as a section of the RAF and moved to Saffraanberg in 1946 taking over the buildings that between the wars had housed the Netherlands-speaking half of the Cadet School.

Two years later a serious fire forced a move to Tongeren and it was not until 1952 that the technical school returned to Saffraanberg.

In September 1957, a Royal Decree established a higher Secundaire Technische School for officers of all three services, and since that time the TSLm has also been responsible for the provision of basic technical training for Army and Navy students destined for the STS.

VERVOLMAKINGS CENTRUM

Insignia: On a blue shield with yellow borders, a dark blue book, green laurels and five red Magisters making coloured smoke (from left to right - white, black, yellow, red and white).
Motto: Nil
Status: Currently flying Magisters and T-33As from their base at Brustem.

The Vervolmakings Centrum is one of three centres subordinate to the Groepering Opleiding en Training and is responsible for all flying instruction.

Following the return of the Belgian Air Force to its homeland in 1946 a training base was set-up at St. Truiden which on 23 May 1947 became the Voortgezette Vliegopleidings School. The Jachtschool followed in 1948, and a specialised instrument training flight, the smaldeel Vliegen Zonder Zicht was established in 1962.

Participation in a joint training scheme with the Dutch from 1961 involved some advanced training being undertaken in Holland, but the replacement of the F-84F by different types in the two countries halted the programme some ten years later.

VOORTGEZETTE VLIEGOPLEIDING SCHOOL

Insignia: An Egyptian falcon, a re-incarnation of the god Horus, over the Roman numeral 'V'. Colouring in natural metal.
Motto: Nil
Status: Currently operated as No's 7 and 9 Smal delen from Brustem, flying Magisters.

Fairey Battles of 5 Verkennings-en-Bommenwerperssmaldeel, III Groep, III Regiment, formed at Evère in 1936, flew in defence of Belgium during May 1940 commanded by Captain Charly de Hepcée.

On 23 May 1947 the falcon insignia of 5 Verkenningssmaldeel was revived at Brustem for the School voor Voortgezette Vliegopleiding (Continuation Flying-training School), whose first commandant was Major Prévot. The 109th course was the first to undergo complete training on home soil flying seventeen Harvards, seven Oxfords and a few Spitfires, but on 21 December 1953 the 126th course left for Kamina in the Congo where the school was re-established with a strength of sixty Harvards.

No sooner had the first replacement Magisters arrived, however, than civil war forced the closure of the school in July 1960, and its return to Brustem, where it opened again on 22 January 1962. Together with the T-33s of the Vlieg Instructie Smaldeel, the Voorgezette Vliegopleiding School constituted the Vlieg Opleidings Wing. An agreement with Holland on 1 November 1961 divided flying training and conversion flying with Magisters remaining at Brustem for advanced instruction and the T-33 transferring to Gilze Rijen. Having received its 45th and last Magister on 10 January 1962, the Voorgezette Vliegopleiding School was integrated into the Vervolmakings Centrum van de Luchtmacht (Central Advanced Training School) combining with the Officieren School on 24 April 1962. The school was divided into two sections in June 1971, adopting the identities of 7 and 9 Smaldelen.

WING TELECOMMUNICATIES

Insignia: A grey tower with white windows above two red lightning flashes; a golden lion in a black shield flanked by two white wings and a black telephone handpiece, all on a blue shield with red borders.
Motto: Nil
Status: Currently based at Evère, no aircraft on charge.

Originally known as the Groep Telecommunicaties, the unit was previously a part of the Onderhouds Eenheid van het Radar Scherm until this was designated 2 Depot on 1 July 1959. When 2 Depot became 22 Wing on 1 October 1966 the Groep was raised in status to a Wing and consists of an Administratief Smaldeel and two groups; the Maintenance Groep, divided between Evère and Batavia and the Groep Exploitatie Telecommunicaties (Working Group).

WING SUPPORT MISSILES

Insignia: In a divided shield of light blue and dark blue, with black edges; a black cogwheel, a yellow flash and a white Nike-Hercules missile with red flames.
Motto: Nil
Status: Currently based at Düren.

With the formation of 9 and 13 Missile Wings the air force also established a maintenance and support unit in 1962, named Steuneenheid Telegeleide Tuigen Grond-Lucht (Air-Ground Missile Support Unit), becoming the Steuneenheid Misseles (Missile Support Unit) in 1964, and the Steunwing Misseles (Missile Support Wing) two years later.

Following the move of 9 Wing from Düren to Grefrath in 1969 the wing established a detachment here in addition to those at the base of each missile squadron. A re-organisation in 1971 rationalised the Wing Support Missile Squadrons into four squadrons, the Onderhoudsmaldeel (Support Squadron), the Uitrustingssmaldeel (Equipment Squadron) and two Dienst Smaldelen (Servicing Squadrons).

ZEEMACHT FLIGHT

See 40 Smaldeel.

7361 MUNITIONS MAINTENANCE SQUADRON USAF

Insignia: A red lion with gold claws and crown over a NATO star; in a light blue disc and yellow edging. White lettering '7361MMS' on a light blue background and white 'Kleine Brogel' on a dark blue background.
Motto: Nil
Status: Currently based at Kleine Brogel

Known until 1972 as Detachment 1 of 36 TFW, the Munitions Maintenance Squadron is attached to 10 Wing to store and maintain the tactical nuclear weapons carried by the Starfighters of 23 and 31 Smaldelen.

Orders of Battle

Locations and equipment of the squadrons of the Belgian Air Force are given at various dates during its history.

1 August 1914

Squadron	Base	Aircraft
1 Smaldeel	Luik	Farman
2 Smaldeel	Namen	Farman
3 Smaldeel	Brasschaat	Farman
4 Smaldeel	Brasschaat	Farman

1 March 1915

1 Smaldeel	Koksijde	Farman
2 Smaldeel	Koksijde	Farman
3 Smaldeel	Koksijde	Voisin
4 Smaldeel	Houtem	Farman
5 Smaldeel	Houtem	Farman
6 Smaldeel	Houtem	Be2C

11 November 1918

1 Smaldeel	Houtem	various
2 Smaldeel	Les Moères	Breguet 14A2
3 Smaldeel	Les Moères	Breguet 14A2
4 Smaldeel	Hondschote	Spad S11
5 Smaldeel	Houtem	Breguet 14A2
6 Smaldeel	Houtem	Spad S11
7 Smaldeel	Houtem	Farman F40
8 Smaldeel	Hondschote	Farman F40
9 Smaldeel	Les Moères	Hanriot HD1
10 Smaldeel	Les Moères	Spad S13
11 Smaldeel	Les Moères	Camel

10 May 1940

1 REGIMENT (ARMY OBSERVATION)

1 Groep	1 Smaldeel	Goetsenhoven	Fox II, Fox III
2 Groep	3 Smaldeel	Goetsenhoven	Fox III
3 Groep	5 Smaldeel	Goetsenhoven	Fox III
4 Groep	7 Smaldeel	Goetsenhoven	Fox VI
5 Groep	9 Smaldeel	Bierset	Renard R31
6 Groep	11 Smaldeel	Bierset	Renard R31

2 REGIMENT (PURSUIT/FIGHTER)

1 Groep	1 Smaldeel	Schaffen	Gladiator
	2 Smaldeel	Schaffen	Hurricane
2 Groep	3 Smaldeel	Nijvel	Fiat CR42
	4 Smaldeel	Nijvel	Fiat CR42
3 Groep	5 Smaldeel	Nijvel	Fox VI
	6 Smaldeel	Nijvel	Fox VI

3 REGIMENT (BOMBER/RECONNAISSANCE)

1 Groep	1 Smaldeel	Evère	Fox III
	3 Smaldeel	Evère	Fox III
2 Groep	9 Smaldeel*	Evère	Battle
	11 Smaldeel*	Evère	Battle
3 Groep	5 Smaldeel	Evère	Battle
	7 Smaldeel	Evère	Fox VI

*In process of formation with only three aircraft each. Units disbanded 10.5.40 and aircraft transferred to 3 Groep.

E.V.S.	Wevelghem Deurne Gosseiles St. Denis Westrem	Avro 504, 626, Morane 236 SV-4B
V.V.S.	Wevelghem Goetsenhoven	Firefly, Fox Breguet XIX, Potez 33, Koolhoven FK56
VervolmakingsSchool	Evère	Firefly, SV-4B Avro 626, Potez 33

1 January 1947

160 Wing	349 Squadron	Bevekom	Spitfire XIV
	350 Squadron	Bevekom	Spitfire XIV
161 Wing	351 Squadron	Florennes	Spitfire XIV
	352 Squadron	Florennes	Spitfire XIV
169 Wing	366 Squadron	Evère	Dakota
	367 Squadron	Evère	Anson, Dominie, Oxford, Proctor
E.V.S.		Schaffen	Tiger Moth
T.T.Flt		Koksijde	Martinet

1 January 1950

1 Wing	4 Smaldeel	Bevekom	Meteor F.4
	10 Smaldeel	Bevekom	Mosquito NF.30
	349 Smaldeel	Bevekom	Meteor F.4
	350 Smaldeel	Bevekom	Meteor F.4
	Hulp Smaldeel	Bevekom	Spitfire XIV
2 Wing	1 Smaldeel	Florennes	Spitfire XIV
	2 Smaldeel	Florennes	Spitfire XIV
	3 Smaldeel	Florennes	Spitfire XIV
15 Wing	20 Smaldeel	Evère	Dakota
	21 Smaldeel	Evère	Anson Dominie Oxford
-	15 Smaldeel	Brasschaat	Auster AOP6
EVS		Schaffen	Tiger Moth
VVS		Brustem	Harvard, Oxford
JachSch.		Koksijde	Spitfire IX
Nav. Sch		Evère	as 21 Smaldeel

1 January 1953

1 Wing	4 Smaldeel	Bevekom	Meteor F.8
	10 Smaldeel	Bevekom	Mosquito NF.30
	11 Smaldeel	Bevekom	Meteor NF.11
	349 Smaldeel	Bevekom	Meteor F.8
	350 Smaldeel	Bevekom	Meteor F.8
	Hulp Smaldeel	Bevekom	Meteor F.4/Spitfire XIV
2 Wing	1 Smaldeel	Florennes	F-84E, F-84G
	2 Smaldeel	Florennes	F-84E, F-84G
	3 Smaldeel	Florennes	F-84G
7 Wing	7 Smaldeel	Chièvres	Meteor F.8
	8 Smaldeel	Chièvres	Meteor F.8
	9 Smaldeel	Chièvres	Meteor F.8
10 Wing	23 Smaldeel	Chièvres	F-84G
	27 Smaldeel	Chièvres	F-84G
	31 Smaldeel	Chièvres	F-84G
15 Wing	20 Smaldeel	Melsbroek	C-119F
	21 Smaldeel	Melsbroek	C-47, DC-4, Dominie, Pembroke
EVS		Goetsenhoven	Tiger Moth, SV-4
VVS		Brustem	Harvard
Jacht Sch		Koksijde	Spitfire IX, T-33A Mosquito
Nav. Sch		Melsbroek	as 21 Smaldeel
	15 Smaldeel	Brasschaat	Auster AOP.6

1 January 1956

1 Wing	4 Smaldeel	Bevekom	Meteor F.8
	10 Smaldeel	Bevekom	Mosquito NF.30
	11 Smaldeel	Bevekom	Meteor NF.11
	349 Smaldeel	Bevekom	Meteor F.8
	350 Smaldeel	Bevekom	Meteor F.8
	Hulp Smaldeel	Bevekom	Meteor F4, Spitfire XIV
2 Wing	1 Smaldeel	Florennes	F-84G/F-84F
	2 Smaldeel	Florennes	F-84G/F-84F
	3 Smaldeel	Florennes	F-84G/F-84F
7 Wing	7 Smaldeel	Chièvres	Meteor F.8
	8 Smaldeel	Chièvres	Meteor F.8
	9 Smaldeel	Chièvres	Meteor F.8
9 Wing	22 Smaldeel	Bierset	F-84G
	26 Smaldeel	Bierset	F-84G
	30 Smaldeel	Bierset	F-84G
10 Wing	23 Smaldeel	Kleine Brogel	F-84G
	27 Smaldeel	Kleine Brogel	F-84G
	31 Smaldeel	Kleine Brogel	F-84G
13 Wing	25 Smaldeel	Brustem	Meteor F.8
	29 Smaldeel	Brustem	Meteor F.8
	33 Smaldeel	Brustem	Meteor F.8
15 Wing	20 Smaldeel	Melsbroek	C-119G
		Melsbroek	C-47, DC-4, Pembroke
—	42 Smaldeel	Brustem	RF-84F
EVS		Goetsenhoven	SV-4
VVS		Kamina	Harvard
Jacht Sch		Koksijde	Meteor, T-33A
T.T. Flt		Sylt	Meteor, Mosquito
Landmacht	15 Smaldeel	Brasschaat	Super Cub
	16 Smaldeel	Butzweilerhof	Super Cub

1 January 1959

1 Wing	11 Smaldeel	Bevekom	CF-100
	349 Smaldeel	Bevekom	CF-100
	350 Smaldeel	Bevekom	CF-100
2 Wing	1 Smaldeel	Florennes	F-84F
	2 Smaldeel	Florennes	F-84F
	3 Smaldeel	Florennes	F-84F
7 Wing	7 Smaldeel	Chièvres	Hunter F.6
	8 Smaldeel	Chièvres	Hunter F.6
9 Wing	22 Smaldeel	Bierset	Hunter F.6
	26 Smaldeel	Bierset	Hunter F.6
10 Wing	23 Smaldeel	Kleine Brogel	F-84F
	27 Smaldeel	Kleine Brogel	F-84F
	31 Smaldeel	Kleine Brogel	F-84F
15 Wing	20 Smaldeel	Melsbroek	C-119G
	21 Smaldeel	Melsbroek	C-47, DC-4, DC-6 Pembroke
	Det.	Kamina	C-119G, Devon
—	42 Smaldeel	Brustem	RF-84F
EVS		Goetsenhoven	SV-4
VVS		Kamina	Harvard
T.T. Flt		Sylt	Meteor
VZZ		Bierset	T-33A
Landmacht	15 Smaldeel	Brasschaat	Super Cub
	16 Smaldeel	Butzweilerhof	Super Cub
	17 Smaldeel	Werl	Super Cub
	18 Smaldeel	Merzbruck	Super Cub

1 January 1965

Wing	Smaldeel	Base	Aircraft
1 Wing	349 Smaldeel	Bevekom	F-104G
	350 Smaldeel	Bevekom	F-104G
	OCU	Bevekom	TF-104G
2 Wing	1 Smaldeel	Florennes	F-84F
	2 Smaldeel	Florennes	F-84F
9 Wing	53 Smaldeel	Kaster	Nike-Hercules
	54 Smaldeel		Nike-Hercules
	55 Smaldeel	Kapellen-Erft	Nike-Hercules
	56 Smaldeel	Grefrath (HQ)	Nike-Hercules
10 Wing	23 Smaldeel	Kleine Brogel	F-104G
	31 Smaldeel	Kleine Brogel	F-104G
13 Wing	50 Smaldeel	Duren (HQ)	Nike-Hercules
	51 Smaldeel	Blankenheim	Nike-Hercules
	52 Smaldeel	Euskirchen	Nike-Hercules
15 Wing	20 Smaldeel	Melsbroek	C-119G
	21 Smaldeel	Melsbroek	C-47, DC-4, DC-6, Pembroke
	40 Smaldeel	Melsbroek	C-119G
EVS		Goetsenhoven	SV-4
VVS		Brustem	Magister
VZZ		Bierset	T-33A
SAR Flt		Koksijde	HSS-1
SRT Flt		Koksijde	HSS-1
Landmacht School Smaldeel		Brasschaat	Super Cub, Alouette II, Do27
	16 Smaldeel	Butzweilerhof	Super Cub, Alouette II, Do27
	17 Smaldeel	Werl	Super Cub, Alouette II
	18 Smaldeel	Merzbruck	Super Cub, Alouette II

1 January 1971

Wing	Smaldeel	Base	Aircraft
1 Wing	349 Smaldeel	Bevekom	F104G
	350 Smaldeel	Bevekom	F104G
	OCU	Bevekom	TF104G
2 Wing	1 Smaldeel	Florennes	F-84F
	2 Smaldeel	Florennes	Mirage 5BA, F-84F
	8 Smaldeel	Florennes	Mirage 5BA, 5BD
3 Wing	VZZ	Bierset	T-33A
	42 Smaldeel	Bierset	RF-84F
9 Wing	53 Smaldeel	Kaster	Nike-Hercules
	54 Smaldeel		Nike-Hercules
	55 Smaldeel	Kapellen-Erft	Nike-Hercules
	56 Smaldeel	Grefrath (HQ)	Nike-Hercules
10 Wing	23 Smaldeel	Kleine Brogel	F-104G
	31 Smaldeel	Kleine Brogel	F-104G
13 Wing	50 Smaldeel	Düren (HQ)	Nike-Hercules
	51 Smaldeel	Blankenheim	Nike-Hercules
	52 Smaldeel	Euskirchen	Nike-Hercules
15 Wing	20 Smaldeel	Melsbroek	C-119G
	21 Smaldeel	Melsbroek	C-47, DC-6, Pembroke
	40 Smaldeel	Melsbroek	C-119G
EVS		Goetsenhoven	SF-260M
VVS		Brustem	Magister
VZZ		Bierset	T-33A
SAR Flt		Koksijde	HSS-1
SRT Flt		Koksijde	HSS-1
Landmacht School Smaldeel		Brasschaat	Alouette II, Do27
	16 Smaldeel	Butzweilerhof	Alouette II, Do27
	17 Smaldeel	Werl	Alouette II
	18 Smaldeel	Merzbruck	Alouette II
Zeemacht		Koksijde	HSS-1

1 January 1977

1 Wing	349 Smaldeel	Bevekom	F 104G
	350 Smaldeel	Bevekom	F-104G
2 Wing	2 Smaldeel	Florennes	Mirage 5BA
	42 Smaldeel	Florennes	Mirage 5BR
3 Wing	1 Smaldeel	Bierset	Mirage 5BA
	8 Smaldeel	Bierset	Mirage 5BD
9 Wing	53 Smaldeel	Kaster	Nike-Hercules
	54 Smaldeel		Nike-Hercules
	55 Smaldeel	Kapellen-Erft	Nike-Hercules
	56 Smaldeel	Grefrath (HQ)	Nike-Hercules
10 Wing	23 Smaldeel	Kleine Brogel	F-104G
	31 Smaldeel	Kleine Brogel	F-104G
13 Wing	50 Smaldeel	Duren (HQ)	Nike-Hercules
	51 Smaldeel	Blankenheim	Nike-Hercules
	52 Smaldeel	Euskirchen	Nike-Hercules
15 Wing	20 Smaldeel	Melsbroek	C-130H
	21 Smaldeel	Melsbroek	B-727, HS748
			Mystère, Merlin
EVS		Goetsenhoven	SF-260M
VVS	7 Smaldeel	Brustem	Magister
	9 Smaldeel	Brustem	Magister
VZZ ——	11 Smaldeel —	Brustem	T-33A
40 Smaldeel	SAR Flt	Koksijde	Sea King
	SRT Flt	Koksijde	HSS-1
	Zeemacht Flt	Koksijde	Alouette III
Landmacht SLV		Brasschaat	Alouette II, Islander
	16 Smaldeel	Butzweilerhof	Alouette II, Islander
	17 Smaldeel	Werl	Alouette II
	18 Smaldeel	Merzbruck	Alouette II

NOTES Each operational Wing consists of the following sections:—

(a) Wing HQ
(b) Two or three squadrons of aircraft
(c) Electron Smaldeel (Electronics squadron)
(d) Smaldeel Maintenance Vliegtuigen (Aircraft Servicing Squadron)
(e) Smaldeel Lijn en Bewapening (Armament Squadron)
(f) Groep Vliegveld Beveiliging (Airfield Defence Group)

c, d and e form the Groepering Maintenance.

The strengths of the Belgian air arms as at 1.1.77 are officially given as:—

	aircraft	helicopters
Airforce	270	8
Navy		4
Army	22	66
Rijkswacht		8

Luchtmacht Organisation 1977

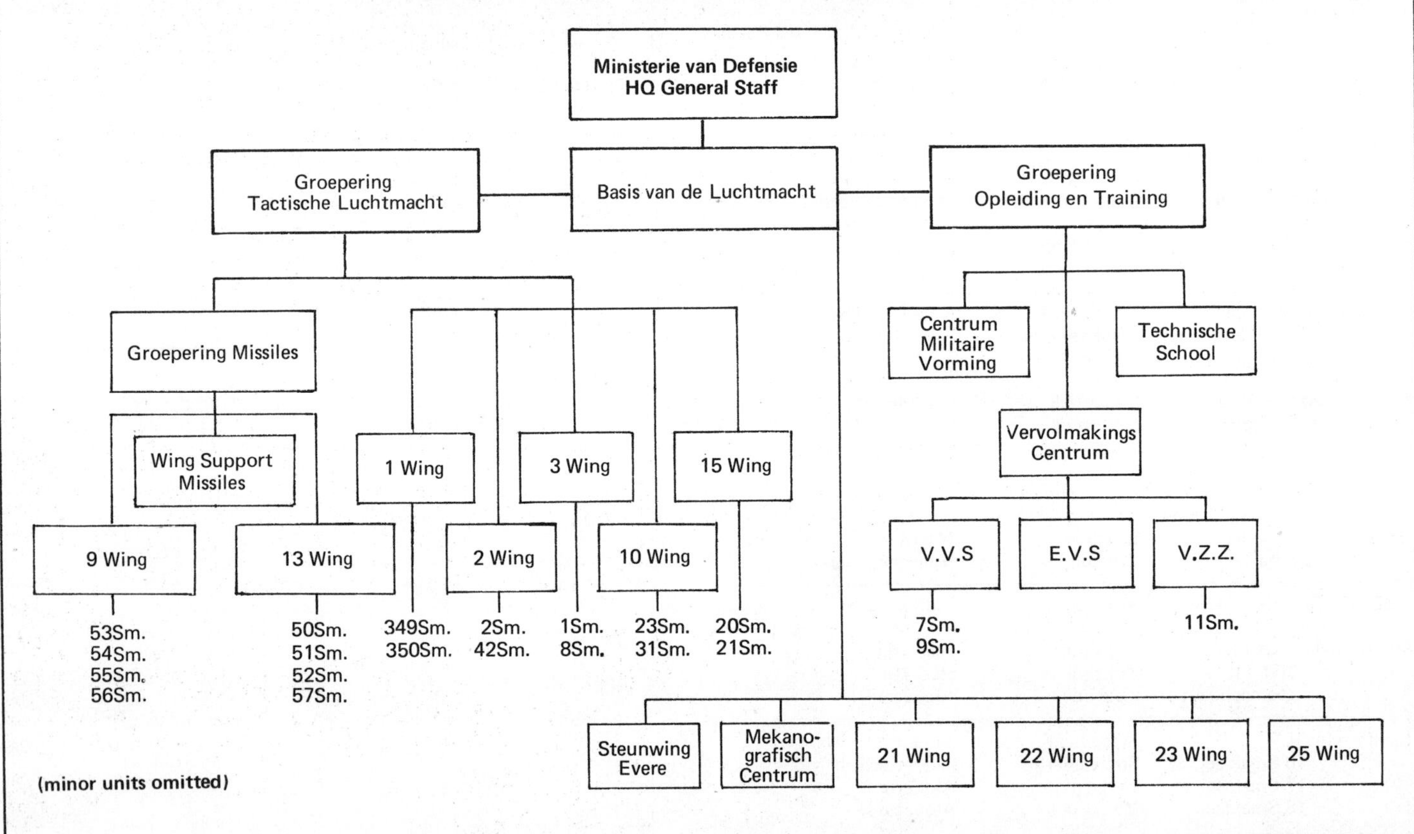

Aero Commander 560F 'OTCWB' was unique in being operated without a serial, but merely with the radio call sign carried at the top of the fin. It was used on Royal flights until sold in 1973. *(via Barry J. Collman, Air-Britain Commander specialist)*

Aerospatiale SA.330C Puma G-03 one of three used by the Rijkswacht, the Belgian equivalent of the French Gendarmerie, but maintained by the Army air corps at Brasschaat. *(B. Marselis)*

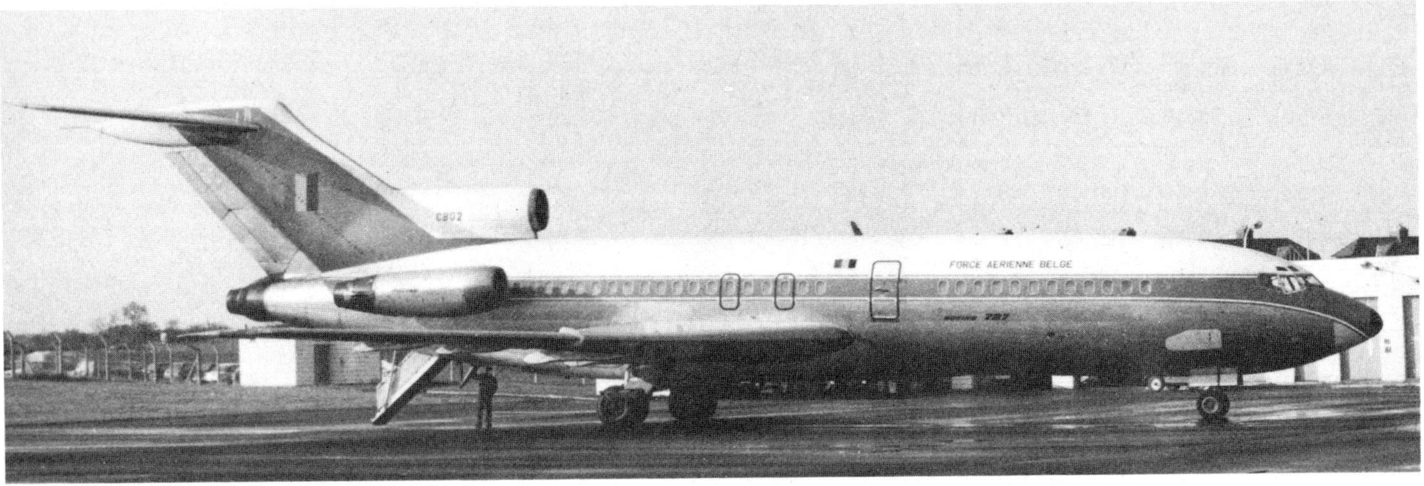
CB02
FORCE AERIENNE BELGE

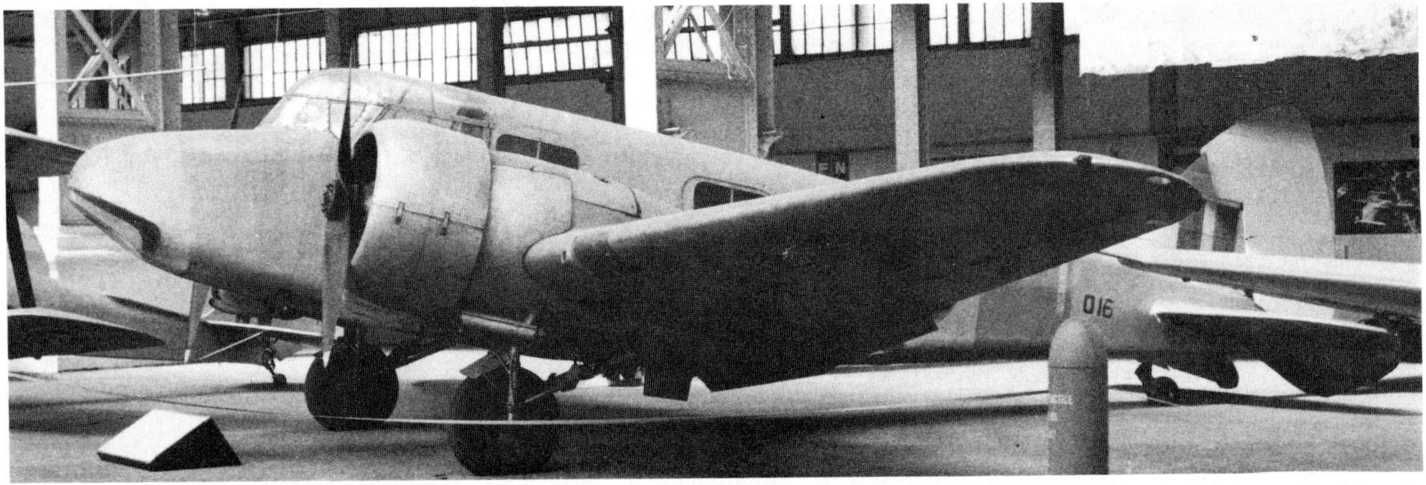
016

Boeing 727-29C CB-02
To replace the ageing DC-6 fleet, 21 Smaldeel now operate two ex-Sabena 727s from Melsbroek, the military side of Brussels National Airport. *(MAP/I. MacFarlane)*

Airspeed Oxford O-16 was delivered in a batch of 42 ex-Royal Air Force aircraft in 1947-49, and is now preserved in the Musée de l'Armée in Brussels. *(Tony Lloyd)*

Auster AOP.6 A-15
Twenty-two new AOP.6 aircraft were received for Army co-operation in 1947. A-15 was ex VT995, and following disposal on the civil market, was eventually returned to military markings for 'gate-guardian' duties at Brasschaat. *(B. Marselis)*

Auster AOP.6 A-16 is now preserved in the Musée de l'Armée. Note the silver overall colours compared with the camouflaged A-15 above. *(MAP/S. Croes)*

Avro Canada CF-100 Canuck Mk.5 AX-16 photographed at Binbrook in April 1963, displays the badge of two 'morning stars' on the air intake, indicating 349 Smaldeel. *(MAP)*

Avro Canada CF-100 Canuck Mk.5 AX-52 visited Coltishall for the 1960 Battle of Britain display. 350 Smaldeel identified its Canucks by an unofficial flash on the air intake; the formal badge, the head of a Belgian warrior, is beneath the windscreen. *(Cambridge Aviation Photographers)*

Avro Canada CF-100 Canuck Mk.5 AX-9 of 11 Smaldeel, seen at Gutersloh in August 1960. *(MAP)*

Dassault Mirage 5BA BA-26 visiting Upper Heyford in July 1976 for the U.S. Bicentennial celebrations shows the insignia of 2 Smaldeel, the second squadron to equip with the ground attack version of the Mirage. *(S.G. Richards)*

Dassault Mirage 5BA BA-44 serves with 1 Smaldeel, and carries that unit's thistle insignia in white on the fin. *(MAP)*

Dassault Mirage 5BA BA-49 carries on its fin the full official thistle badge within a circular surround of 1 Smaldeel. *(via A.D. Annis)*

Dassault Mirage 5BA BD-02
Sixteen two-seat 5BD models were delivered for use by 8 Smaldeel, the Mirage operational conversion unit. BD-02 has been 'zapped' by AKG 51 of the Luftwaffe, and carries that unit's owl badge on the nose. *(via A.D. Annis)*

Dassault Mirage 5BDs BD-04 and BD-06 seen at Mildenhall in October 1972 illustrate clearly that each Mirage in Belgian service carries its own unique camouflage pattern. *(S.G. Richards)*

Dassault Mirage 5BD BD-05 seen at Woodbridge in August 1976. *(Cambridge Aviation Photographers)*

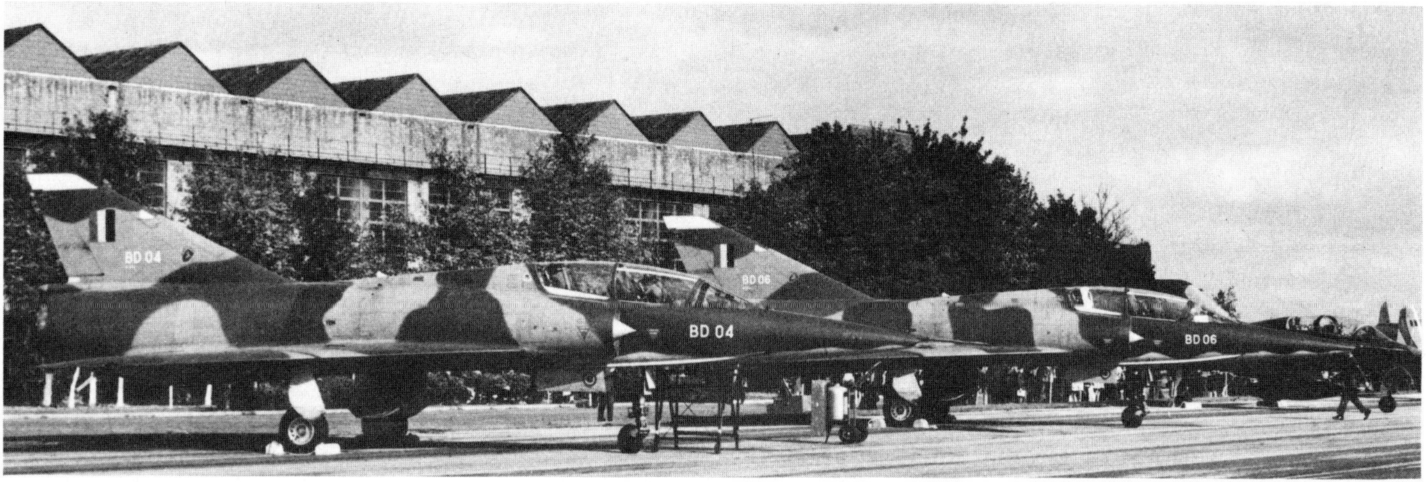
BD 04
BD 04
BD 06
BD 06

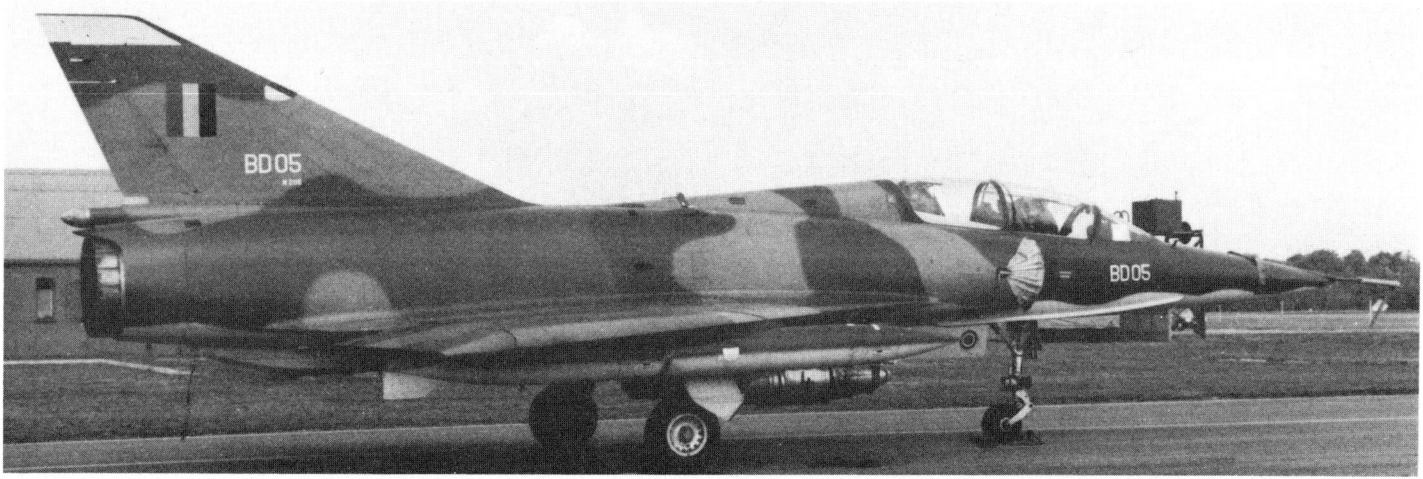
BD05
BD05

Dassault Mirage 5BD BD-04 seen at Bevekom in June 1972 is unusual in displaying the 8 Smaldeel insignia of a folded-paper bird in a shield immediately forward of the fin serial. *(P. Glas)*

Dassault Mirage 5BR BR-01 was the Dassault-built prototype of the reconnaissance version of the Mirage, and originally serialled MR-01. 42 Smaldeel's 'devil' badge is carried on the fin. *(E. Ragas)*

Part Three

Aircraft Review

Markings and colour schemes

As it was established in 1946 almost entirely with ex-RAF aircraft, the Luchtmacht adopted British practices in marking of aircraft, most noticeably in the perpetuation of a coding system.

Only 349 and 350 Smaldelen received *bona fide* RAF codes, the remainder were 'invented' as new units formed. Codes used were:

Code	Belgian unit	Previous RAF user
B2	Target Towing Flight	Nil
EB	30 Smaldeel	41 Squadron
GE	349 Smaldeel	349 and 58 Squadron
GV	Hulp Smaldeel	103 Squadron; 1652 HCU
H8	42 Smaldeel	Nil
IQ	Jagd School	150 Squadron
IS	22 Smaldeel	Nil
JE	26 Smaldeel	195, 198 and 610 Squadron
KT	10 Smaldeel	47 and 457 Squadron and 53 OTU
K5	33 Smaldeel	Pocklington Station Flight
MN	350 Smaldeel	350 Squadron
MS	29 Smaldeel	23 and 273 Squadron; Hornet Conv. Flight
ND	11 Smaldeel	236 Squadron
OV	8 Smaldeel	197 Squadron
RA	27 Smaldeel	100 and 410 Squadron
SV	4 Smaldeel	218 Squadron and 1663 CU
S2	9 Smaldeel	Nil
VR	2 Smaldeel	419 Squadron
VT	25 Smaldeel	84 and 216 Squadron and 1556 Flt.
XO	24 Smaldeel	112 Squadron and 57 OTU.
YL	3 Smaldeel	Nil
Z6	23 Smaldeel	Nil
3R	1 Smaldeel	3RSU
7J	7 Smaldeel	Nil
8S	31 Smaldeel	Nil

Transport aircraft carried the full radio call sign, commencing 'OT':

Radio Callsign	Aircraft type	Unit
OTAEA – OTAEF	C-119G	20/40 Smaldeel
OTCAA – OTCAT	C-119G	40 Smaldeel
OTCBA – OTCBT	C-119G	20 Smaldeel
OTCDA – OTCDF	DC-6	21 Smaldeel
OTCWB	Aero Commander	21 Smaldeel
OTCWU – OTCWV	DC-4	21 Smaldeel
OTZAA – OTZAL	Pembroke	21 Smaldeel
OTZCC	Anson	21 Smaldeel
OTZCJ	Dominie	21 Smaldeel
OTZKA – OTZKC	Sycamore	15 Wing
OTZKD – OTZKP	HSS-1	Heli Smaldeel

Further information concerning radio call-signs not marked on the aircraft is contained in the individual aircraft lists, and in Appendix E.

During the early fifties, aircraft served in natural metal finish with black lettering and codes. This lettering changed to white when camouflage was introduced, and under-wing serials were also deleted. Codes were discontinued in the early sixties and replaced on fighter aircraft by a repeat of the whole serial in small lettering, or the last two or three in larger numerals.

Camouflage has now spread to some training aircraft, eg SIAI SF.260 and T-33A. The former, like all recent deliveries having black numbering and no fuselage roundels. The Hercules has taken this one step further and omits the fin flash. Roundels are the national colours of black (centre), yellow and red.

Squadron badges are worn by most aircraft, usually on the tailfin. In some instances, eg 1 and 10 Wings, aircraft are pooled and carry the wing badge in lieu. In the F-84 era, units were additionally identified by differing coloured bands behind the air intake. Serial number prefixes are dealt with elsewhere.

Serial numbers

Luchtmacht: The post war method of allocating serials to aircraft followed the pattern used previously, with each type being given a one or two letter prefix followed by a number normally beginning at '1'. Aircraft marks were often differentiated by use of other prefix letters, whilst in the case of the Dakota, the installation of special fitments such as photographic equipment or VIP interiors added extra letters to the prefix.

First allocations of single letter prefixes mainly followed the initial letter of the aircraft type, or its function, *viz*—

A	Auster AOP.6
C	Airspeed Consul
D	Dominie
G	Magister
H	Harvard
K	Dakota
O	Oxford
P	Proctor
R	Miles Martinet (Remorquer – target tug)
T	DH.82A Tiger Moth

Others, delivered at the same time, used two letters to identify marks

MA	Mosquito TT.3
MB	Mosquito NF.30
MC	Mosquito TT.6
NA	Anson 1
NB	Anson 2
SG	Spitfire XIV (Griffon)
SM	Spitfire IX (Merlin)

Because the letter 'M' was in use when the first Meteors were delivered, the type 'E' was allocated, and as the aircraft was a fighter the Meteor F.4s began at EF-1. The Meteor F.8 used the next letter after 'F' and so started from EG-1 and the trainer mark T.7 adopted 'ED'. Night Fighter variants used, appropriately, 'EN'. Four further types used the single letter system before its abandonment.

B	Sycamore HR.14B
C	Chipmunk
D	Dove
V	Stampe SV.4

During the early fifties, the first letter of the prefix was adopted as a functional classification in addition to denoting particular types, as with the Meteor.

Transport aircraft employed the American-type 'C' or occasionally 'K'

CB	Boeing 727
CF	Fairchild-Swearingen Merlin
CH	Hercules
CM	Mystere 20
CP	Boxcar (Packet)
CS	HS.748
KX	DC-4
KY	DC-6

Fighters used 'F' with the exception of the Hunter, classified as an Interceptor, and CF-100 which was All-weather.

FC	TF-104G
FR	RF-84F
FS	F-84E
FU	F-84F
FX	F-104G
FZ	F-84E/G
ID	Hunter F.4
IF	Hunter F.6
AX	CF-100

But T continued as a second letter identification for trainers.

FT	T-33A (Fighter –trainer)
MT	Magister
ST	SIAI SF.260M

The trainer rule was broken in 1970 when twin seat Mirages appeared as 'MD', the 'D' being the Dassault designation for a trainer, as was 'A' for strike and 'R' for reconnaissance. These prefixes soon changed to BA, BD and BR, reflecting the full Dassault mark number.

One-off uses by the Luchtmacht are RM for Pembroke C.51, LB for Super Cub L-21, RS Sea King and the re-allocation of 'B' to the HSS-1 commencing at B-4, following three Sycamores.

Landmacht: When the army took charge of its aviation element on 1 May 1954, the 'Licht Vliegwezen van de Landmacht' (Army Air Corps) adopted a system of serialling consisting of the prefix 'OL–' denoting 'Observation Leger' (light observation), a letter indicative of the aircraft type and a number commencing '01'. Aircraft used have been:

A	Alouette II
B	Islander
D	Do.27
G	Puma (for Rijkswacht)
L	Super Cub L-18

Three of the first Alouettes delivered were despatched to the Congo, where they began a series commencing at '51', and those delivered later for the Rijkswacht started with '91'. The 'OL' prefix was deleted in 1974, and replaced by the application of call-signs to helicopters in small lettering on the nose.

Zeemacht: The Navy, having originally been allocated two of the Luchtmacht HSS-1 helicopters, received three of their own Alouette III in 1970–71. These were serialled in the 'M' range.

Individual aircraft service histories

In this section are listed all the aircraft used by the Luchtmacht since 1945, in alphabetical order of manufacturers. Whilst we have attempted to make the information as complete as possible, readers will understand that for some of the earlier types, the available material is very scarce.

AERO COMMANDER 560F

One example of this twin engined high-wing light transport was delivered on 10 September 1961, and operated by 21 Smaldeel for Royal flights. It was unusual in that no serial number was allocated; instead it carried its radio callsign OTCWB on the fintip in small lettering. C/n was 560F-1069-25, and the aircraft was sold as F-BTYZ in April 1973.

AEROSPATIALE SA.330C PUMA

The Puma was designed as a heavy assault helicopter for Army use, and first flew in 1965. Principal user is the French Army, with 140, but worldwide civil and military sales now total over 400.
Three aircraft were delivered to Brasschaat in 1973 for Rijkswacht use. At that time they had the 'OL—' prefix to their serials, but these were removed in 1974.

C/n	Serial	Callsign
1225	G01	OTGIA
1237	G02	OTGIB
1265	G03	OTGIC

AIRSPEED CONSUL

The Consul was a refurbished version of the Oxford (see below) used for communications. The Belgian Air Force had four aircraft based in the Congo, from 1948.

C/n	Serial	Ex-	Sold as
5190	C-31		ZS-DNJ
5178	C-32	NM593/G-AKCV	
5189	C-33		ZS-DNK
5133	C-34	X6859/G-AJLO	ZS-DNL

AIRSPEED OXFORD

Oxfords served the RAF throughout World War 2 as radio trainers, blind approach trainers and communications aircraft, with large numbers disposed of during 1946-50. These included 42 aircraft for 367 Squadron, and later 21 Smaldeel, with serials O-1 to O-42, all withdrawn from use circa 1954. Known ex-RAF aircraft are detailed below —

RAF serial		C/n	Serial
V3775	*	3257	
DF523	*		O-1
EB796	*		
LX728	*	3423	O-29
MP301	*		O-6, later G-AOUT
MP430	*		O-27, later OO-DUC
NM694	+	3825	
NM696		3827	
NM702		3833	
NM720		3851	
NM736		3852	
NM777	*	3878	
PH351	*	4170	
PH352	*	4171	
PH415	+	4223	
PH417	+	4225	
PH425	+	4233	
PH459		4246	
PH460	+	4247	
PH461	+	4248	

** delivered March-November 1947*
+ delivered via Rollasons, Croydon, Nov 1950

AUSTER AOP.6

Following the Marks 1 to 5 developed by Auster from pre-war Taylorcraft designs, the AOP.6 light observation aircraft entered production in the closing stages of the war. The RAF took 274, Canada 30, and South Africa five. Twenty two aircraft were delivered to Belgium from new production for the RAF, and allocated serials at random A-1 to A-22. The first delivery was on 8 August 1947, and the last on 14 January 1948, all aircraft going to 369 Smaldeel, later 15 Smaldeel. The type was replaced by the Piper Super Cub during 1954-55, and the survivors civilianised. In the listing below, only the known tie-ups are shown, the blanks in the serial column representing the balance of the serials from A-1 to A-22.

C/n	RAF serial	Serial	Became
2815	VT976		
2816	VT977		
2817	VT978	A-9	OO-FDD
2818	VT979	A-3	OO-FDA
2819	VT980		
2820	VT981	A-7	OO-FDB
2821	VT982		OO-OVL
2822	VT983		
2823	VT984		
2828	VT985		
2829	VT986		
2830	VT987		
2824	VT988	A-8	OO-FDC
2825	VT989	A-13	OO-FDG
2826	VT990	A-11	OO-FDE
2827	VT991	A-12	OO-FDF
2831	VT992	A-18	OO-FDK
2832	VT993	A-17	OO-FDJ
2833	VT994		
2834	VT995	A-15	OO-FDH
2835	VT996	A-16	OO-FDI
2836	VT997	A-22	OO-FDL

AVRO ANSON

The first of over 11,000 Ansons was delivered to the RAF in 1935 for evaluation, and the type was used for maritime patrol during the early months of the war. Training and communications versions were in production up to 1952.
Fifteen aircraft from RAF surplus stocks were delivered to Belgium during 1946-47 for use as communications aircraft by 367 Smaldeel, later 21 Smaldeel, until replaced by Percival Pembrokes in 1954.

Anson I serials NA-1 to NA-13
Anson 12 serials NB-1 and NB-2

Ex-RAF serials included PH697 and PH767 delivered September 1946, and N5028, DG773, EG208, EG268, LT126, LT605 and NK619 (NA-6) all delivered May 1947. Known aircraft written off are NA-11 on 27.1.48 and NA-13 on 18.8.55.

AVRO CANADA CF-100 CANUCK Mk.5

This twin-jet all-weather fighter was produced in Canada between 1950 and 1959, all of the 692 aircraft with the exception of the Belgian examples being for the Royal Canadian Air Force, who withdrew most of the type in 1965.
Fifty-three aircraft were built between February 1957 and April 1958, and delivered for operation by 1 Wing (11, 349 and 350 Smaldelen). The official transfer date for all aircraft was 7 October 1957, with the first aircraft entering service on 17 December 1957, and the last delivery on 1 July 1958. Most withdrawn early to mid-sixties; the aircraft marked 'K' in the list below were noted at Koksijde awaiting scrapping in 1964, and those marked 'H' sold as scrap to a Mr. Hevghen.

Serial	ex-RCAF	Remarks
AX-1	18685	H
AX-2	18686	H
AX-3	18687	H
AX-4	18688	H
AX-5	18689	H
AX-6	18690	w/o 5.1.61

AX-7	18691	K
AX-8	18692	K
AX-9	18693	H
AX-10	18694	K
AX-11	18695	H
AX-12	18696	H
AX-13	18697	w/o 6.7.62
AX-14	18698	
AX-15	18699	w/o 14.4.61
AX-16	18700	K
AX-17	18701	H
AX-18	18702	K
AX-19	18703	H
AX-20	18704	H
AX-21	18705	K
AX-22	18706	K
AX-23	18707	H
AX-24	18708	K
AX-25	18709	H
AX-26	18710	H
AX-27	18711	H
AX-28	18712	H
AX-29	18713	K
AX-30	18714	w/o 9.8.60
AX-31	18715	H
AX-32	18716	w/o 27.9.60
AX-33	18717	H
AX-34	18718	w/o 30.7.59
AX-35	18719	K
AX-36	18720	H
AX-37	18721	K
AX-38	18722	K
AX-39	18723	H
AX-40	18724	K
AX-41	18725	w/o 13.3.63
AX-42	18726	K
AX-43	18727	K
AX-44	18728	K
AX-45	18729	K
AX-46	18730	K
AX-47	18732	K
AX-48	18733	K
AX-49	18734	K
AX-50	18735	K
AX-51	18736	K
AX-52	18737	K
AX-53	18738	K

It is possible that 18731, which was Belgian according to Canadian records, crashed before delivery, and that the original order was for 54 aircraft, i.e three squadrons of eighteen.

BOEING 727-29C

Two examples of Boeing's highly successful tri-jet transport were transferred from the Sabena fleet in 1975-76 as replacements for the DC-6s with 21 Smaldeel; they had been with Sabena since mid-1967 when they were built.

C/n	Serial	Ex-	Callsign
19402	CB-01	OO-STB	BAF21
19403	CB-02	OO-STD	BAF22

BRISTOL SYCAMORE HR.14B

Britain's first indignenous production helicopter found large orders from the RAF, Luftwaffe, and Australian Navy. Three Belgian aircraft were acquired for SAR work in the Congo, but used initially at Filton for crew-training. B-1 and B-3 were flown to the Congo by C-119 in June 1954 and based at Baka with 15 Wing; B2 was delivered to Melsbroek and then Baka in early 1955.

c/n	serial	Ex-	Call-sign	Fate
13199	B-1	G-AMWP	OTZKA	w/o 31.3.60
13200	B-2	G-AMWR	OTZKB	Abandoned 1960
13201	B-3	G-AMWS	OTZKC	w/o 20.1.58

DASSAULT MIRAGE 5

The Mirage 5 series was developed by Dassault as a simplified-avionics (and thus cheaper) version of the very popular Mirage III. First flight was in May 1967, and an order was received from Israel, though this was later vetoed by the French Government. The Belgian order came after comparative studies with the SAAB Draken and A-4 Skyhawk, and costs were further reduced by the implementation of a licence-building agreement, and the use of American electronics. 106 aircraft ordered 25 August 1968, to replace the F-84F and RF-84F, and all except three, built by SABCA/Avions Fairey.

Mirage 5BA

Dassault-built prototype MA-01 first flown 6 March 1970 and delivered 29 June 1970, later re-serialled BA-01. All delivered in natural metal finish and later camouflaged, each aircraft individually with a slightly different pattern. Mark 5BA is the strike version.

C/n	Serial	Unit	Remarks
01	BA-01	2Sm	w/o 2.77
02	BA-02	2Sm	
03	BA-03	1Sm	ex 2Sm
04	BA-04	2Sm	
05	BA-05	1Sm	ex 2Sm
06	BA-06		w/o 19.11.76
07	BA-07	1Sm	
08	BA-08	1Sm	ex 2Sm
09	BA-09	1Sm	
10	BA-10	1Sm	ex 2Sm
11	BA-11	2Sm	
12	BA-12	1Sm	w/o 5.12.75
13	BA-13	2Sm	
14	BA-14	2Sm	w/o 15.11.75*
15	BA-15	1Sm	ex 8Sm
16	BA-16	2Sm	
17	BA-17		
18	BA-18	2Sm	
19	BA-19	1Sm	
20	BA-20	2Sm	
21	BA-21	1Sm	
22	BA-22		
23	BA-23	1Sm	
24	BA-24	2Sm	
25	BA-25		w/o 28.3.72
26	BA-26	2Sm	
27	BA-27	2Sm	
28	BA-28	1Sm	ex 2Sm
29	BA-29	1Sm	
30	BA-30	1Sm	
31	BA-31	1Sm	
32	BA-32	1Sm	w/o 6.9.73
33	BA-33	1Sm	
34	BA-34	1Sm	ex 2Sm
35	BA-35	1Sm	
36	BA-36	1Sm	w/o 4.74
37	BA-37	1Sm	
38	BA-38	1Sm	w/o 6.9.73
39	BA-39	1Sm	
40	BA-40	1Sm	
41	BA-41	1Sm	ex 2Sm
42	BA-42	1Sm	
43	BA-43	2Sm	
44	BA-44	1Sm	
45	BA-45	1Sm	
46	BA-46	1Sm	
47	BA-47	1Sm	w/o 28.6.76
48	BA-48	1Sm	
49	BA-49	1Sm	w/o 5.6.75*
50	BA-50	2Sm	
51	BA-51	1Sm	ex 2Sm
52	BA-52	2Sm	
53	BA-53	2Sm	
54	BA-54		
55	BA-55	1Sm	ex 8Sm
56	BA-56	1Sm	
57	BA-57	2Sm	ex 1Sm
58	BA-58	2Sm	w/o 10.1.77
59	BA-59	1Sm	
60	BA-60		
61	BA-61	1Sm	
62	BA-62	1Sm	
63	BA-63	1Sm	

**W/os marked thus are confirmed crash dates, and serials, but the tie-ups are not confirmed.*

Mirage 5BD

A prototype two-seat trainer was built by Dassault and serialled MD-01, but later re-serialled BD-01. The first SABCA-built Mirage was BD-02 delivered to the Luchtmacht on 28 August 1970.

C/n	Serial	Unit	Remarks
201	BD-01	8Sm	
202	BD-02	8Sm	w/o 14.7.75
203	BD-03	8Sm	
204	BD-04	8Sm	
205	BD-05	8Sm	
206	BD-06	8Sm	
207	BD-07	8Sm	
208	BD-08	8Sm	
209	BD-09	8Sm	
210	BD-10	8Sm	
211	BD-11	8Sm	
212	BD-12	8Sm	
213	BD-13	8Sm	

214	BD-14	8Sm	
215	BD-15	8Sm	
216	BD-16		w/o 7.12.73

Mirage 5BR

Twenty-seven reconnaissance versions of the Mirage 5 to replace RF-84F Thunderflashes with 42 Smaldeel. Prototype MR-01 built by Dassault, later re-serialled BR-01.

C/n	Serial	Unit	Remarks
301	BR-01	42Sm	w/o 2.8.73
302	BR-02	42Sm	w/o 25.3.73
303	BR-03	42Sm	
304	BR-04	42Sm	
305	BR-05	42Sm	w/o 4.4.75
306	BR-06	42Sm	w/o 15.6.73
307	BR-07	42Sm	
308	BR-08	42Sm	
309	BR-09	42Sm	
310	BR-10	42Sm	
311	BR-11	42Sm	w/o 2.5.75
312	BR-12	42Sm	
313	BR-13	42Sm	
314	BR-14	42Sm	
315	BR-15	42Sm	
316	BR-16	42Sm	
317	BR-17	42Sm	
318	BR-18	42Sm	
319	BR-19	42Sm	
320	BR-20	42Sm	
321	BR-21	42Sm	
322	BR-22	42Sm	
323	BR-23	42Sm	
324	BR-24	42Sm	
325	BR-25	42Sm	
326	BR-26	42Sm	
327	BR-27	42Sm	

DASSAULT MYSTERE 20E

Two examples of Dassault's executive jet transport were delivered 24 March and 18 May 1973 for VIP work by 21 Smaldeel.

C/n	Serial	Unit	Callsign
276	CM-01	21Sm	BAF31
278	CM-02	21Sm	BAF32

DE HAVILLAND DH.82A TIGER MOTH

Large numbers of the famous Tiger Moth were produced immediately before and during World War 2, the final British-built example flying in July 1944. Production was undertaken additionally in Canada, New Zealand, Australia, Norway, Portugal and Sweden.
Thirty-one aircraft, ex-RAF post-war, saw Belgian service, originally serialled ETA-1 to ETA-31, and later re-serialled T-1 to T-31. The principal user was the EVS, Goetsenhoven.

Serials T-1 to T-15 allocated to BTS, Snailwell in February 1945, and transferred to Belgian charge on 19 November 1946:

Ex-RAF	Serial	C/n	Fate
DF124	T-1	85873	To OO-EVA
DF152	T-2	85901	Crash 5.8.56
NL926	T-3	86369	Crash 27.7.56
NL934	T-4	86377	Crash 8.54
NL916	T-5	86359	Crash 28.4.52
NL933	T-6	86376	Crash 15.8.51
DF179	T-7	85915	Crash 5.8.55
DF198	T-8	85934	To OO-EVB
DF151	T-9	85900	To OO-EVK
*	T-10		Crashed
*	T-11		Crash 6.6.50
*	T-12		w/o
DF178	T-13	85914	To OO-EVF
*	T-14		Crash 3.12.53
NL891	T-15	86334	To OO-EVG

** The four aircraft thus shown are ex DE776/85690, DF208/85944, NL932/86375 and NL972/86404, but the order is uncertain.*

Serials T-16 to T-30 allocated to BTS, January-April 1946, but never officially transferred:

Ex-RAF	Serial	C/n	Fate
DE418	T-16	85426	Crash 26.11.52
DE721	T-17	85651	To OO-EVC
NM194	T-18	86502	To OO-EVL
NM199	T-19	86507	To OO-EVM
DF201	T-20	85937	Crash 17.7.48
NM207	T-21	86515	To OO-EVP
NM209	T-22	86517	To OO-EVD
DF212	T-23	85948	To OO-EVQ
PG614	T-24	86523	Wfu 1955
EM722	T-25	85953	To OO-EVE
EM744	T-26	85975	Wfu 1956
DE972	T-27	85872	To OO-EVH
EM738	T-28	85969	To OO-EVI
R4771	T-29	82712	To OO-EVJ
DF135	T-30	85884	Wfu 1950

Serial T-31 - replacement aircraft DF126/85875 delivered from RAF stocks 24 June 1947. Crashed 23.3.50, but repaired with parts of K4276, and returned to service, until sold as OO-EVS.

Note re Tiger Moth constructor's numbers. In 1941 production was transferred to Morris Motors at Oxford, this firm assembling T5611-39, 5669-5718, 5749-5788, 5807-13 and from T5891 onwards, perpetuating the de Havilland c/n system. Aircraft in the NL and PG serial ranges were the subject of a direct order with Morris (Contract 2714) rather than transferred production, and whilst documentation quotes these aircraft as c/ns 86172 to 86632, first hand examination suggests that they received Morris c/ns MCO/DH/4249 to 4709. Thus subtraction of the number 81923 from the hypothetical 86.... c/n will give the Morris fuselage number.

DE HAVILLAND DH.89 DOMINIE

The DH.89 Dragon Rapide biplane transport first flew in 1934, with RAF deliveries commencing a year later. During the war the type served as a radio trainer and communications aircraft as the DH.89 Dominie.
Seven aircraft were delivered to Belgium in September 1946 ex-RAF for 367 Squadron (later re-numbered 21 Smaldeel) for communications duties. Two of these were later sold to the COGEA anti-aircraft co-operation firm.

Serial	ex-RAF	C/n	Fate	Callsign
D-1	NR805	6881		OTZCH
D-2)*	NF874	6853		
D-3)*	NF868	6739		
D-4	NR686	6785	To OO-ARN	
D-5	NR688	6787	To OO-ARI, G-APBN and F-OBIA	
D-6	NR776	6852		OT-ZCJ
D-7	NR777	6853		

**serial tie-ups uncertain*

DE HAVILLAND DH.98 MOSQUITO

The private venture Mosquito 'wooden wonder' first flew on 25 November 1940 and was immediately recognised as an outstanding military aircraft, adaptable as a bomber, reconnaissance aircraft, or night fighter. Many examples were presented to Allied air forces after the war and continued in service well into the fifties. Belgian examples consisted of thirty nine aircraft all ex-RAF, in four marks:

Mosquito T/TT.3

Grounded T.3 LR528/6318M delivered to Technische School van de Luchtmacht in July 1947. Seven TT.3 target-towing aircraft delivered July 1947 to February 1948, and used at Koksijde -

Serial	ex-RAF
MA-1	VR333
MA-2	VR335
MA-3	VR338
MA-4	VR339
MA-5	VR341
MA-6	VR342
MA-7	VR343

Mosquito FB.6/TT.6

Instructional FB.6 used in England from June 1945 as NS857/5265M.
Three FB.6 converted to TT.6 October 1951 by Fairey at Ringway, Manchester -

Serial	ex-RAF
MC-1	TE614
MC-2	TE663
MC-3	TE771

Mosquito NF.17

One instructional aircraft at Tongeren, delivered October 1947 - HK327/6343M.

Mosquito NF.30

Twenty-two aircraft delivered for 10 and 11 Smaldelen between January 1948 and January 1949, with a further two, MB-23 and 24, following in 1953 -

Serial	ex-RAF	Remarks
MB-1	NT446	Scrapped 1956
MB-2	MM768	Scrapped 1956
MB-3	NT322	w/o 9.6.49
MB-4	NT300	Scrapped 1956
MB-5	NT368	w/o 5.12.52
MB-6	MT465	w/o 15.6.49
MB-7	MV559	w/o 25.11.49
MB-8	MT491	Instructional
MB-9	NT314	w/o 14.10.49
MB-10	MT499	w/o 28.4.52
MB-11	NT377	Scrapped 1956
MB-12	NT384	Scrapped 1956
MB-13	NT317	w/o 30.9.51
MB-14	NT362	w/o 25.8.50
MB-15	NT375	Scrapped 1956
MB-16	NT387	Scrapped 1956
MB-17	NT501	Scrapped 1956
MB-18	MM757	w/o 14.10.50
MB-19	NT275	w/o 5.12.52
MB-20	NT330	Scrapped 1956
MB-21	MM687	Scrapped 1956
MB-22	NT332	Scrapped 1956
MB-23	RK935	Crashed on delivery
MB-24	RK952	Preserved

In addition NT450 and NT563 were used for instruction at Tongeren.

DE HAVILLAND DH.104 DOVE

The Dove twin-engined transport began production in 1945, and when the line was closed over twenty years later over 540 had been built.
Twelve aircraft were operated by the Force Publique du Congo Belge in Luchtmacht markings; the first three were delivered 1949-50, seven in 1952-54, and the final two in November 1959. At least six were taken over by the Katangan Air Force during the uprising in July 1960, and four of these were re-possessed by the Congolese Air Force in 1963 and re-serialled 9T-P40 to P43.

Serial	C/n	Ex-	Fate
D-10	04252	'CGG'*	w/o
D-11	04080	'CGG'*	To KAT-11, w/o 8.61
D-12	04052	G-AJZU OO-CFC	w/o 7.10.58
D-13	not used		
D-14	04367		To KAT-14
D-15	04442		To KAT-15, KA-T15 9Q-COB
D-16	04443		To KAT-16 Abandoned 7.60
D-17	04447		To KAT-17 w/o 8.61
D-18	04154	VP-YHV	To KAT-18
D-19	04013	OO-AWE OO-CWE	To KAT-19 and w/o
D-20	04103	OO-CFE	w/o
D-21	04506		Abandoned 7.60
D-22	04507		To KAT-22, KA-T22, abandoned 1963

** 'CGG' = Congo Governor General; 04080 was the first 'CGG', replaced by 04252.*

DE HAVILLAND DH.114 HERON 2

The four-engined 14-seat development of the Dove. One Belgian aircraft c/n 14055, ex G-ANPV not taken up, delivered ex UK 25 May 1954 as 'CGG' replacing the previous Doves with the Congo Governor General. Seized by the Katangans in July 1960, becoming KAT-01, but reported captured and destroyed at Elizabethville in August 1961.

DE HAVILLAND CANADA DHC.1 CHIPMUNK

The Canadian company's first product was intended as a Tiger Moth replacement in the training role, and was very successful. Most European examples were built under licence in the U.K. or Portugal, but the Belgian aircraft originated from the Canadian production line.
Two aircraft delivered for evaluation in 1948, in competition with the Stampe SV-4 (which gained the order as Tiger replacement), and retained until sold to the Royal Antwerp Aero Club on 6.3.56.

C/n	Serial	Fate
19	C-1	To OO-PHS w/o 19.9.70
20	C-2	To OO-MER

DORNIER Do 27D-J1

The Dornier 27 was designed in Spain in the fifties, but the main production was 428 for the German armed forces, with whom the type still serves. Civilian and export sales totalled over 100, including 12 for Belgium. First Belgian delivery was a batch of three in 1960, and the type served with 15 and 16 Smaldelen until replaced in 1976 by the FBN Islander.

C/n	Serial	Callsign	Unit	Fate
2057	OL-D01	OTAMA	15Sm	Scrapped
2058	OL-D02	OTAMB	16Sm	Scrapped
2059	OL-D03	OTAMC	16Sm	Scrapped
2101	OL-D04	OTAMD	16Sm	Preserved
2102	OL-D05	OTAME	15Sm	
2103	OL-D06	OTAMF	15Sm	
2104	OL-D07	OTAMG		
2105	OL-D08	OTAMH		
2106	OL-D09	OTAMI	16Sm	Scrapped
2107	OL-D10	OTAMJ	15Sm	
2108	OL-D11	OTAMK	16Sm	Scrapped
2109	OL-D12	OTAML	15Sm	Scrapped

DORNIER-DASSAULT-BREGUET ALPHA JET

First flown on 26 October 1973 the Alpha Jet was the winner of a Franco-German competition for an advanced trainer. Four prototypes and six pre-production aircraft precede 200 aircraft each for France and Germany, with orders also from Turkey and Belgium.
The Belgian aircraft are sixteen on order, plus seventeen more on option; they are due to replace the Lockheed T-33A with 11 Smaldeel, and the Magister with the VVS. Serials may possibly be AT-1 to AT-33.

DOUGLAS C-47 DAKOTA

The ubiquitous and apparently ageless Dakota began its Belgian service with the delivery of a batch of fifteen aircraft between July and November 1946. A further 25 arrived February 1947 to February 1950. with the forty-first and final aircraft on 19 February 1951. Twenty four returned to the USAF 1952-53 for forwarding to other European air forces, and the final two aircraft remaining in service K-8 and K-10 were officially retired 26 January 1976, though K-8 was noted still in use during the summer of 1976!
Role prefixes included 'FC' for reconnaissance, 'P' for photography, and 'R' VIP, but these later removed in favour of the simple 'K'.

Serial	Ex-USAF/ RAF	C/n	Callsign/ Disposal
K-1	43-49240	15056/ 25601	OTCWA To OO-SMA
KP-2	43-16275	20741	OTCWB w/o 13.9.55

K-3	43-48306	14122/ 25567	OTCNA French AF 1952
KP-4	43-16398	20864	OTCNB Preserved 1971
K-5	43-16413	20879	OTCWC Neth. AF 1952 X-2
KP-6	43-16418	20884	OTCNC w/o 15.6.57
K-7	43-48485	14301/ 25746	OTCWD To USAF 1952
KFC-8	43-48495	14311/ 25756	OTCND 21Sm 'BAF71'
K-9	43-48585	14401/ 25846	OTCNE French AF 1952
KR-10	43-48590	14406/ 25851	OTCWE Wfu 1.76 'BAF72'
K-11	43-48608	14424/ 25869	OTCWF French AF 1952
K-12	43-48619	14435/ 25880	OTCNC French AF 1952
K-13	43-48785	14601/ 26046	OTCNG Wfu 1965
K-14	43-48896	14712/ 26107	OTCNH w/o 1947
K-15	43-48396	14212/ 25607	OTCNF French AF 1952
K-16	43-16357	20823	OTCWG wfu 1972
K-17	43-48789	14605/ 26050	OTCNJ French AF 1952
K-18	43-48787	14603/ 26048	OTCWH w/o 1.12.58
K-19	44-76225	15810/ 32557	OTCWI Katanga 'KAT-03'
K-20	43-49608 KK178	15424/ 26869	OTCNK French AF 1952
K-21	44-76226 KN304	15810/ 32558	OTCWJ Katanga 'KAT-02'
K-22	43-49604 KK174	15522/ 26805	OTCNL French AF 1953
K-23	43-48990 KJ959	14806/ 26251	OTCNM French AF 1952
K-24	43-49824 KN219	16540/ 27085	OTCNN French AF 1952
K-25	43-48927 KN218	15643/ 27088	OTCWK USAF 1952
K-26	44-76300 KN325	15884/ 32632	OTC . . Neth. AF 1952 X-3
K-27	43-49735 KN341	15551/ 26996	OTC . . Neth. AF 1952 X-4
K-28	44-76332 KN347	15916/ 32664	OTC . . OO-SNC 1951
K-29	44-76717 KN512	16301/ 33049	OTC . . French AF 1952
K-30	43-49950 KN276	15766/ 27211	OTCNQ Sold
K-31	44-76486 KN391	16064/ 32812	OTCNR Scrapped
K-32	44-76991 KN637	16575/ 33233	OTCNS Neth. AF 1952 X-5
K-33	44-77068 KN690	16652/ 33400	OTCNT USAF 1952
K-34	44-76835 KP218	16419/ 33167	OTCNU USAF 1952
K-35	44-77101 KN555	16685/ 33433	OTCWN French AF 1952
K-36	44-77116 KP229	16700/ 33448	OTCWO USAF 1952
K-37	44-77069 KN689	16653/ 33401	OTCWP French AF 1952
K-38	43-48555 KN877	14371/ 25816	OTCWQ French AF 1952
K-39	44-76423 KN381	16007/ 32755	OTCWR French AF 1952
K-40	44-76912 KN601	16496/ 33244	OTCWS Wfu 1967
K-41	43-48670 KJ900	14486/ 25931	OTCWT w/o 13.9.51

DOUGLAS DC-4

Douglas' first four-engined airliner was produced in large numbers during the war as the C-54, with civil production continuing for a time thereafter.
Two aircraft were purchased by Belgium for flights to the Congo, operated by 21 Smaldeel, until retired in 1969.

KX-1 callsign OTCWU c/n 10336 ex 42-72221, Bu39173, NC49539, OO-CBS, delivered 2.10.50. Retired 1969, and delivered to a cafe at Torhout in 1973.

KX-2 callsign OTCWV c/n 42987 ex OY-DFO, OO-SBL, delivered 5.6.54. Sold 1969 as G-AZSI.

DOUGLAS DC-6

Four examples of this post-war development of the DC-4 saw Belgian service, all with 21 Smaldeel on transport duties. The first two were DC-6A freighters originally ordered by Slick Airlines, but delivered instead to Belgium on 25.5.58, whilst the second pair were earlier examples which had been delivered as DC-6As to Sabena in 1954, but converted on arrival to DC-6Cs. They were always owned by the Belgian Government, and transferred to the air force on 10.10.60. The type was replaced by the Boeing 727, and all four were stored at Koksijde late 1976.

C/n	Serial	Ex-	Callsign
45458	KY-1	N7819C*	OTCDA
45518	KY-2	N7820C	OTCDB
44420	KY-3	OO-CTO	OTCDE
44421	KY-4	OO-CTP	OTCDF

** operated by Sabena for the Ministry of National Defence as OO-SMR 21.5.59 to 10.59.*

Note: The registrations F-BYCG, 'CH, 'CI, CJ were reserved by Secmafer of Nice in Feb. 77 for KY-1, KY-2, KY-3, KY-4 respectively.

FAIRCHILD C-119F/G FLYING BOXCAR

The twin-boom freighter C-119 appeared in 1947 as a development of the C-82 Packet. The type served with the USAF until replaced in the mid-fifties by the Lockheed Hercules, the same type which eventually supplanted the C-119 with the Belgian Air Force during 1972-73. Forty-six aircraft were delivered from 1952 onwards, initially C-119F (1951 serials), and later C-119G. Eight of the C-119Fs were passed to the Norwegian Air Force June-September 1956, and the remaining ten then converted to 'G' standard in 1959, with some re-serialling. The C-119Gs CP-19 to CP-40 were delivered between 10.8.53 and 20.3.54, with CP-41 to CP-46 following ex-USAF in February 1958. All aircraft served with 15 Wing only, amassing some 154, 157 flying hours until retired 1972-73. Most went into storage at Koksijde, but CP-29 and 37 are believed sold to Ethiopia, and CP-46 is with the Brussels Museum.

Serial	Ex-USAF	Callsign	Remarks
CP-1	51-2692	OTCAA	To Norway
CP-2	51-2693	OTCAB	To Norway
CP-3	51-2694	OTCAC	To CP-14(2)
CP-4	51-2695	OTCAD	To Norway
CP-5	51-2696	OTCAE	To CP-11(2)
CP-6	51-2697	OTCAF	To Norway
CP-7	51-2698	OTCAG	To Norway
CP-8	51-2699	OTCAH	To Norway
CP-9	51-2700	OTCAI	
CP-10	51-2701	OTCAJ	
CP-11	51-2702	OTCAK	To Norway
CP-11	(2) – see CP-5 above		
CP-12	51-2703	OTCAL	
CP-13	51-2704	OTCAM	
CP-14	51-2705	OTCAN	To Norway
CP-14	(2) – see CP-3 above		
CP-15	51-2706	OTCAO	
CP-16	51-2707	OTCAP	
CP-17	51-2690	OTCAQ	
CP-18	51-2691	OTCAR	
CP-19	52-6034	OTCAS	
CP-20	52-6032	OTCAT	
CP-21	52-6022	OTCBA	

CP-22	52-6023	OTCBB	
CP-23	52-6021	OTCBC	w/o 12.12.61
CP-24	52-6039	OTCBD	
CP-25	52-6042	OTCBE	w/o 12.12.61
CP-26	52-6028	OTCBF	
CP-27	52-6026	OTCBG	
CP-28	52-6038	OTCBH	
CP-29	52-6047	OTCBI	
CP-30	52-6027	OTCBJ	
CP-31	52-6035	OTCBK	
CP-32	52-6045	OTCBL	
CP-33	52-6032	OTCBM	
CP-34	52-6050	OTCBN	
CP-35	52-6052	OTCBO	
CP-36	52-6044	OTCBP	w/o 19.7.60
CP-37	52-6055	OTCBQ	
CP-38	52-6051	OTCBR	
CP-39	52-6046	OTCBS	
CP-40	52-6058	OTCBT	
CP-41	53-7829	OTAEA	
CP-42	53-7843	OTAEB	
CP-43	53-8130	OTAEC	
CP-44	53-8141	OTAED	
CP-45	53-8143	OTAEE	w/o 26.6.63
CP-46	53-8151	OTAEF	

FAIREY BRITTEN-NORMAN BN.2A-21 ISLANDER

The Islander, an outstanding example of the success of free enterprise in Britain, first flew in 1965, and its simplicity, adaptability and STOL performance have won hundreds of orders world wide. Following the acquisition of Britten-Norman by Fairey the main production was switched to Gosselies, where the twelve Belgian aircraft were built, though they were ferried to Bembridge for fitting and painting. Deliveries were between May 1976 and January 1977, and the type replaced the Dornier 27 with 16 Smaldeel and with the School van het Licht Vliegwezen.

C/n	serial	Callsign	Unit	Ex-
466	B-01	OTALA	SLV	G-BDJA
468	B-02	OTALB	16Sm	G-BDHG
476	B-03	OTALC	SLV	G-BDJV
498	B-04	OTALD	16Sm	G-BDPN
504	B-05	OTALE	16Sm	G-BDPP
510	B-06	OTALF	16Sm	G-BDPU
523	B-07	OTALG	SLV	G-BDVX
531	B-08	OTALH	SLV	G-BDZI
533	B-09	OTALI	16Sm	G-BDZK
541	B-10	OTALJ	SLV	G-BEDW
549	B-11	OTALK	SLV	G-BEED
553	B-12	OTALL	16Sm	G-BEFI

FOUGA CM.170R MAGISTER

The V-tailed Magister jet trainer first flew in 1952, and 588 were built in France, plus 194 under licence in Germany, and smaller numbers in Finland and Israel.
Forty-five Fouga-built aircraft were delivered between 29 January 1960 and 10 January 1962 as Harvard replacements, initially to Kamina , and later to Brustem with 7 and 9 Smaldelen of the VVS. MT-37, 38 and 39 were originally c/ns 294-296, but these were transferred to the Congolese Air Force, and replaced on the production line by later examples, MT-40 to 45 similarly c/ns 297-302 for Katanga but placed in storage; 298 and 299 to Irish Air Corps in 1976.
Five further examples, all Messerschmitt-built examples retired by the Luftwaffe, were delivered in 1969 (MT-46 to MT-50), and followed by nine refurbished aircraft which replaced aircraft previously written off, adopting both the serials and c/ns of the earlier aircraft!
In the list below VVS/RD indicates 'Rode Duivels'.

C/n	Serial	Unit	Remarks
258	MT-1	VVS	
259	MT-2	VVS	w/o 1958
'259'	MT-2	VVS	
260	MT-3	VVS	
261	MT-4	VVS	
262	MT-5	VVS/RD	
263	MT-6	VVS/RD	w/o 2.76
264	MT-7	VVS	w/o 29.9.67
'264'	MT-7	VVS	
265	MT-8	VVS	
266	MT-9	VVS	w/o 27.2.70
267	MT-10	VVS	w/o 10.3.65
'267'	MT-10	VVS	
268	MT-11	VVS/RD	
269	MT-12	VVS/RD	
270	MT-13	VVS	
271	MT-14	VVS	
272	MT-15	VVS/RD	
273	MT-16	VVS	
274	MT-17	VVS	w/o 25.9.69
'274'	MT-17	VVS	
275	MT-18	VVS/RD	
276	MT-19	VVS	w/o
277	MT-20	VVS	w/o 19.6.68
'277'	MT-20	VVS	
278	MT-21	VVS/RD	
279	MT-22	VVS	w/o
'279'	MT-22	VVS	
280	MT-23	VVS/RD	
281	MT-24	VVS	
282	MT-25	VVS	w/o 10.63
'282'	MT-25	VVS	
283	MT-26	VVS	
284	MT-27	VVS	
285	MT-28	VVS	w/o
286	MT-29	VVS	w/o 2.64
'286'	MT-29	VVS	
287	MT-30	VVS	w/o 2.3.70
'287'	MT-30	VVS	
288	MT-31	VVS/RD	
289	MT-32	VVS/RD	
290	MT-33	VVS	
291	MT-34	VVS	
292	MT-35	VVS	
293	MT-36	VVS	
312	MT-37	VVS	
313	MT-38	VVS	w/o
314	MT-39	VVS	w/o
317	MT-40	VVS	
318	MT-41	VVS	w/o
319	MT-42	VVS	
322	MT-43	VVS	
323	MT-44	VVS	
327	MT-45	VVS	w/o
145	MT-46	VVS	ex AA237, ED392
203	MT-47	VVS	ex Luftwaffe
204	MT-48	VVS	ex AA293, AA204
222	MT-49	VVS	ex Luftwaffe
224	MT-50	VVS	ex AA240

Other aircraft previously with the 'Rode Duivels' have been MT-2 (the first), 3, 7, 9, 16, 20, 22 (first), 27, 29, 35.
MT-2 (No.2), MT-7 (No.2), MT-9 (No.2), MT-10 (No.2), MT-17 (No.2), MT-20 (No.2), MT-22 (No.2) and MT-25 (No.2) are no longer on charge and were presumably only on loan.

GENERAL DYNAMICS F-16A/B

Flown for the first time on 20 January 1974, the F-16 has been chosen in preference to the Northrop F-17 as the USAF's next air combat fighter. Orders so far total: 650 U.S.A., 84 Netherlands, 102 Belgium, 48 Denmark and 72 for Norway. General Dynamics will build the USAF aircraft incorporating some European components, whilst the construction of the other 348 will be split 174 each between Fokker and SABCA. However, whilst Fokker will produce aircraft from its own and imported components, Fairey SA will handle all F-16 assembly in Belgium, leaving only final assembly and flight test for SABCA. USAF deliveries will commence in August 1978, those from SABCA in January 1979, and from Fokker in June 1979.

The Belgian order for 102 aircraft was placed on 7 June 1975, and the following delivery schedule is anticipated —

	F-16A	F-16B
1979	14	4
1980	9	3
1981	17	2
1982	20	2
1983	21	1
1984	9	—

The F-16A will replace the Starfighter with the four squadrons of 1 and 10 Wings, with the twin-seat F-16B attached to the training flights of these units. A further fourteen aircraft are on option.

GLOSTER METEOR

Britain's first operational jet fighter entered RAF service in 1944, and in its later F.4, T.7 and F.8 versions saw world-wide sales. Four marks of Meteor saw Belgian service -

Meteor F.4

Forty-eight new aircraft ordered March 1949 and built by Glosters. Used by 349 and 350 Smaldeel from June 1949 until replaced by the F.8 in 1951, whereupon twenty were converted by Avions Fairey at Gosselies to T.7, and others to Israel. During the course of sales demonstrations in May 1947, Gloster's company demonstrator G-AIDC had crashed at Melsbroek.

C/n	Serial	Fate
G5/246	EF-1	w/o 28.8.52
G5/247	EF-2	w/o 20.12.52
G5/248	EF-3	w/o 9.11.54
G5/249	EF-4	To ED-13
G5/250	EF-5	w/o 29.9.54
G5/251	EF-6	
G5/252	EF-7	To ED-14
G5/253	EF-8	
G5/254	EF-9	To ED-15
G5/255	EF-10	To ED-16
G5/256	EF-11	To ED-17
G5/257	EF-12	To ED-18
G5/258	EF-13	To ED-19
G5/259	EF-14	w/o 19.11.53
G5/260	EF-15	To ED-20
G5/261	EF-16	To ED-21
G5/262	EF-17	w/o 10.10.51
G5/263	EF-18	To ED-22
G5/264	EF-19	To ED-23
G5/265	EF-20	
G5/266	EF-21	To ED-24
G5/267	EF-22	w/o 30.6.52
G5/268	EF-23	To ED-25
G5/269	EF-24	w/o 29.7.54
G5/270	EF-25	To ED-26
G5/271	EF-26	w/o 2.2.53
G5/272	EF-27	w/o 20.11.53
G5/273	EF-28	To ED-27
G5/274	EF-29	
G5/275	EF-30	To ED-28
G5/276	EF-31	
G5/277	EF-32	To ED-29
G5/278	EF-33	
G5/279	EF-34	
G5/280	EF-35	w/o 20.12.51
G5/281	EF-36	
G5/282	EF-37	
G5/283	EF-38	
G5/284	EF-39	To ED-30
G5/285	EF-40	w/o 31.3.50
G5/286	EF-41	To ED-31
G5/287	EF-42	To ED-32
G5/288	EF-43	w/o 23.6.52
G5/290	EF-44	
G5/291	EF-45	
G5/292	EF-46	w/o 15.2.51
G5/293	EF-47	w/o 15.4.55
G5/294	EF-48	

Meteor T.7

Forty-three aircraft used principally by the Jachtschool, and acquired in several batches:

Three newly-built Gloster examples delivered 1 Wing late 1948 -

C/n	Serial
G5/212	ED-1
G5/213	ED-2
G5/214	ED-3

Nine unused ex-RAF delivered as shown -

ex-RAF	Serial	Delivered
WF817	ED-4	3.7.51
WF827	ED-5	3.7.51
WF818	ED-6	3.7.51
WL399	ED-7	25.7.52
WL428	ED-8	8.4.53
WL427	ED-9	25.2.53
WL415	ED-10	13.4.53
WH171	ED-11	13.4.53
WH174	ED-12	13.4.53

Twenty produced by Avions Fairey from surplus F.4 (as above), fitted with Gloster-built noses, and including some with F.8 type tails -

Ex-	Serial	Remarks
EF-4	ED-13	
EF-7	ED-14	
EF-9	ED-15	
EF-10	ED-16	
EF-11	ED-17	
EF-12	ED-18	
EF-13	ED-19	
EF-15	ED-20	
EF-16	ED-21	w/o 3.10.55
EF-18	ED-22	
EF-19	ED-23	
EF-21	ED-24	
EF-23	ED-25	
EF-25	ED-26	
EF-28	ED-27	
EF-30	ED-28	
EF-32	ED-29	
EF-39	ED-30	w/o 23.2.56
EF-41	ED-31	
EF-42	ED-32	

Second allocation, of unused aircraft ex RAF –

Ex-RAF	Serial	Delivered	Remarks
WL486	ED-33	1.9.53	
WN320	ED-34	3.9.53	
WS140	ED-35	18.9.53	w/o 19.5.54
WS141	ED-36	25.1.54	w/o 20.8.54
XF273	ED-37	25.1.54	

Third allocation, of used aircraft, ex RAF –

Ex-RAF	Serial	Delivered	Remarks
WA688	ED-38	9.10.56	
WH114	ED-39	9.11.56	
WA684	ED-40	30.11.56	
WH117	ED-41	30.11.56	
WG970	ED-42	29.1.57	
WF814	ED-43	8.2.57	

Meteor F.8

240 aircraft from various sources delivered from 1951 to 7 Wing (7, 8 and 9 Smaldelen), 1 Wing (4, 349, and 350 Smaldelen), 13 Wing (25,29 and 33 Smaldelen), and 5 Wing (24 Smaldeel), the latter used for target-towing.

Fokker-built aircraft, delivered December 1950 to February 1954 –

C/n	Serial	Remarks
6315	EG-1	
6316	EG-2	
6317	EG-3	
6318	EG-4	
6319	EG-5	
6320	EG-6	
6321	EG-7	
6322	EG-8	
6323	EG-9	
6324	EG-10	
6329	EG-11	
6332	EG-12	
6333	EG-13	
6335	EG-14	
6336	EG-15	
6337	EG-16	
6338	EG-17	
6339	EG-18	
6343	EG-19	
6344	EG-20	
6348	EG-21	
6349	EG-22	
6350	EG-23	
6351	EG-24	
6352	EG-25	
6355	EG-26	
6356	EG-27	
6357	EG-28	
6358	EG-29	
6361	EG-30	
6362	EG-31	
6363	EG-32	
6364	EG-33	
6389	EG-34	
6390	EG-35	
6391	EG-36	
6392	EG-37	
6393	EG-38	
6394	EG-39	
6378	EG-40	
6435	EG-41	
6436	EG-42	
6437	EG-43	
6438	EG-44	
6439	EG-45	
6440	EG-46	
6443	EG-47	
6444	EG-48	
6445	EG-49	
6446	EG-50	
6447	EG-51	
6448	EG-52	
6449	EG-53	
6450	EG-54	
6451	EG-55	
6452	EG-56	
6453	EG-57	
6454	EG-58	
6460	EG-59	
6461	EG-60	

6474	EG-61
6480	EG-62
6481	EG-63
6482	EG-64
6483	EG-65
6484	EG-66
6494	EG-67
6502	EG-68
6503	EG-69
6504	EG-70
6505	EG-71
6506	EG-72
6507	EG-73
6508	EG-74
6509	EG-75
6513	EG-76
6514	EG-77
6520	EG-78
6521	EG-79
6522	EG-80
6523	EG-81
6524	EG-82
6528	EG-83
6529	EG-84
6530	EG-85
6531	EG-86
6532	EG-87
6533	EG-88
6534	EG-89
6541	EG-90
6542	EG-91
6543	EG-92
6544	EG-93
6545	EG-94
6546	EG-95
6547	EG-96
6548	EG-97
6549	EG-98
6554	EG-99
6555	EG-100
6561	EG-101
6562	EG-102
6563	EG-103
6564	EG-104
6569	EG-105
6570	EG-106
6571	EG-107
6572	EG-108
6573	EG-109
6574	EG-110
6580	EG-111
6581	EG-112
6582	EG-113
6583	EG-114
6584	EG-115
6591	EG-116
6592	EG-117
6593	EG-118
6594	EG-119
6598	EG-120
6599	EG-121
6600	EG-122
6601	EG-123
6602	EG-124
6603	EG-125
6604	EG-126
6608	EG-127
6609	EG-128
6610	EG-129
6611	EG-130
6612	EG-131
6613	EG-132
6614	EG-133
6622	EG-134
6623	EG-135
6624	EG-136
6632	EG-137
6633	EG-138
6634	EG-139
6639	EG-140
6640	EG-141
6641	EG-142
6642	EG-143
6643	EG-144
6644	EG-145

Ex-RAF aircraft delivered via Fokker October 1951 –

Ex-RAF	Serial	Remarks
WF692	EG-146	
WF693	EG-147	
WF691	EG-148	
WF701	EG-149	
WH448	EG-150	

Thirty aircraft assembled by Avions Fairey from Fokker-built components –

C/n	Serial	Remarks
6455	EG-151	
6456	EG-152	
6457	EG-153	
6458	EG-154	
6459	EG-155	
6475	EG-156	
6476	EG-157	
6477	EG-158	
6478	EG-159	
6479	EG-160	
6495	EG-161	
6496	EG-162	
6497	EG-163	
6498	EG-164	To OO-ART
6499	EG-165	
6515	EG-166	
6516	EG-167	
6517	EG-168	
6518	EG-169	
6519	EG-170	
6535	EG-171	
6536	EG-172	
6537	EG-173	
6538	EG-174	
6539	EG-175	
6556	EG-176	
6557	EG-177	
6558	EG-178	To OO-ARV
6559	EG-179	
6560	EG-180	

Fairey allocated c/ns 7001 onwards to Meteor F.8 aircraft which they assembled, though whether this sequence began with EG-151, or EG-224 below, is unclear.

Serials EG-181 to EG-200 were not allocated.

Twenty-three ex-RAF aircraft, delivered between 14 November 1950 and 26 January 1951 –

Ex-RAF	Serial	Remarks/Codes
VZ450	EG-201	MN-E, B2-E
*	EG-202	
*	EG-203	
*	EG-204	
VZ457	EG-205	
VZ459	EG-206	MN-L, B2-T
WA883	EG-207	
WA870	EG-208	B2-W
WA878	EG-209	
WA876	EG-210	
VZ553	EG-211	
*	EG-212	
WA884	EG-213	
WA887	EG-214	
WA888	EG-215	
WA889	EG-216	B2-Q
*	EG-217	
WA895	EG-218	
WA898	EG-219	B2-V
WA755	EG-220	B2-S
WA900	EG-221	B2-U
WA901	EG-222	
WA902	EG-223	

** the aircraft thus marked were ex WA881, VZ499, VZ562, WA892 and VZ566. Though the serial tie-ups are unconfirmed, these are believed to be in the order in which they fill the gaps above.*

Thirty-seven aircraft assembled by Avions Fairey from Gloster-built parts –

Serial	Remarks/Codes
EG-224	K5-K
EG-225	B2-H
EG-226	
EG-227	
EG-228	
EG-229	B2-C
EG-230	
EG-231	
EG-232	
EG-233	
EG-234	B2-O
EG-235	B2-N
EG-236	
EG-237	'VAN'
EG-238	B2-J
EG-239	5K-D, B2-Z
EG-240	
EG-241	5K-W, B2-F
EG-242	
EG-243	
EG-244	MN-A
EG-245	B2-P
EG-246	
EG-247	B2-R
EG-248	
EG-249	
EG-250	
EG-251	
EG-252	
EG-253	B2-M
EG-254	
EG-255	B2-I
EG-256	

EG-257
EG-258
EG-259
EG-260

EG-261 to 270 were intended as above, but were cancelled.

Meteor NF.11

Twenty-four Armstrong-Whitworth built aircraft, ex-RAF, for 10 and 11 Smaldelen, the first twelve delivered July-September 1952, and the remainder January-March 1956. Those surviving in service were replaced by the CF-100 Canuck and withdrawn in August 1958, and most civilianised for operation by COGEA from Koksijde.

Ex-RAF	Serial	Codes	Remarks
WD726	EN-1	ND-L, KT-L	w/o 26.8.55
WD775	EN-2	KT-E	To OO-ARR
WD777	EN-3	ND-G, KT-G	w/o 1.12.53
WD728	EN-4		w/o 12.7.54
WD729	EN-5	KT-S	To OO-ARW
WD730	EN-6	KT-W	To OO-ARO
WD727	EN-7	ND-X, KT-X	w/o 27.4.53
WD731	EN-8	KT-Y	w/o 12.4.56
WD732	EN-9	KT-Z	w/o 25.11.54
WD733	EN-10	KT-U	w/o 1.8.57
WD735	EN-11	KT-T	To OO-ARX
WD736	EN-12		
WD602	EN-13	ND-A	
WD724	EN-14	ND-B	w/o 5.12.56
WD622	EN-15	ND-C	w/o 19.1.56
WD594	EN-16	ND-D	To OO-ARQ
WD760	EN-17	ND-E	w/o 30.10.57
WD661	EN-18	ND-F	To OO-ARS
WD590	EN-19	ND-G	To OO-ARZ
WD596	EN-20	ND-H	To OO-GEV
WD741	EN-21	ND-J	To OO-ARP
WD763	EN-22	ND-K	w/o 13.3.56
WM221	EN-23	ND-L	To OO-GEZ
WM263	EN-24		w/o 12.12.56

HAWKER HURRICANE II

By 1945 the famous Hurricane was already obsolete, having been replaced by the Tempest and Typhoon with the RAF for operational use. However, some remained on charge for a few years for miscellaneous duties.
Three aircraft from the Metropolitan Communications Squadron were passed to Belgium for use by 367 Squadron —
LF345 delivered 2 June 1946 "ZA-P"
PZ754 delivered 4 September 1946 "RA-A"
PZ769 delivered 2 September 1946 "ZC"
The latter was reserialled as 'THS5Z' with Technische Hogeschool Saffraanberg as a communications aircraft in airworthy condition. The others retained RAF serials, and LF345 RAF roundels.
RAF records suggest that a fourth aircraft, LF658, was delivered 2.9.46.

HAWKER HUNTER

The Hunter was the RAF's principal trans-sonic fighter of the fifties, the prototype having flown on 20 July 1951. The first of the mass production variants was the F.4, followed in 1955 by the higher-powered F.6. The Hunter remained in front-line service with the RAF until 1971 and is still employed in the training role.
In Belgium both F.4s and F.6s were used, all being European-built under licence, and using American offshore funding. The F.4s had a very short service life, 92 being rebuilt as F.6s. The later models were withdrawn from 1962, and many sold to Hawkers for remanufacture.

Hunter F.4

One hundred and twelve aircraft delivered 1956-57 to 7, 9 and 13 Wings. Withdrawn from use 1957, and 92 aircraft rebuilt as Mark F.6. ID-1 to ID-64 built by SABCA and Avions Fairey; the Fairey-built aircraft were allocated c/ns from 8001 onwards, but it is still unclear whether this sequence includes aircraft assembled from Fokker-made parts. ID-101 to 148 built by Fokker and delivered September 1956 to May 1957.

C/n	Serial	Remarks
	ID-1	
	ID-2	
	ID-3	
	ID-4	
	ID-5	
	ID-6	
	ID-7	
	ID-8	
	ID-9	
	ID-10	
	ID-11	
	ID-12	
	ID-13	
	ID-14	
	ID-15	
	ID-16	
	ID-17	
	ID-18	
	ID-19	
	ID-20	
	ID-21	
	ID-22	
	ID-23	
	ID-24	
	ID-25	
	ID-26	
	ID-27	
	ID-28	
	ID-29	
	ID-30	
	ID-31	
	ID-32	
	ID-33	
	ID-34	
	ID-35	
	ID-36	
	ID-37	
	ID-38	
	ID-39	
	ID-40	
	ID-41	
	ID-42	
	ID-43	
	ID-44	
	ID-45	
	ID-46	
	ID-47	
	ID-48	
	ID-49	
	ID-50	
	ID-51	
	ID-52	
	ID-53	
	ID-54	
	ID-55	
	ID-56	
	ID-57	
	ID-58	
	ID-59	
	ID-60	
	ID-61	
	ID-62	
	ID-63	
	ID-64	
8650	ID-101	
8651	ID-102	
8654	ID-103	
8655	ID-104	
8656	ID-105	
8657	ID-106	
8658	ID-107	
8659	ID-108	
8662	ID-109	
8663	ID-110	
8664	ID-111	
8665	ID-112	
8666	ID-113	
8667	ID-114	
8670	ID-115	
8671	ID-116	
8672	ID-117	
8673	ID-118	
8674	ID-119	
8675	ID-120	
8678	ID-121	
8683	ID-122	
8679	ID-123	
8680	ID-124	
8681	ID-125	
8682	ID-126	
8686	ID-127	
8691	ID-128	
8687	ID-129	
8688	ID-130	
8689	ID-131	
8690	ID-132	
8695	ID-133	
8696	ID-134	
8697	ID-135	

8698	ID-136
8703	ID-137
8704	ID-138
8705	ID-139
8706	ID-140
8710	ID-141
8711	ID-142
8712	ID-143
8713	ID-144
8718	ID-145
8719	ID-146
8720	ID-147
8721	ID-148

Hunter F.6

One hundred and forty four aircraft all built by Fokker and assembled by Avions Fairey (59), SABCA (29) and Fokker (56). Deliveries between 19 June 1957 and 2 December 1958 to 7 and 9 Wings.
Those aircraft which had not been written off or scrapped at the end of their service were sold to Hawkers for remanufacture and re-sale and delivered to Dunsfold from 30 October 1962 in British Class B markings. These marks are given in the 'Remarks' column below, and further details of their subsequent histories are in the table following the serial list.

C/n	Serial	Remarks
8745	IF-1	To G-9-187
8746	IF-2	To G-9-107
8747	IF-3	
8748	IF-4	To G-9-124
8749	IF-5	
8750	IF-6	To G-9-70
8751	IF-7	To G-9-154
8752	IF-8	To G-9-127
8753	IF-9	To G-9-133
8754	IF-10	To G-9-71
8755	IF-11	To G-9-97
8756	IF-12	
8757	IF-13	To G-9-128
8758	IF-14	To G-9-72
8759	IF-15	
8760	IF-16	To G-9-123
8761	IF-17	To G-9-158
8762	IF-18	To G-9-148
8763	IF-19	Parts to G-APUX
8764	IF-20	To G-9-73
8765	IF-21	To G-9-74
8766	IF-22	To G-9-108
8767	IF-23	
8768	IF-24	To G-9-75
8769	IF-25	To G-9-162
8770	IF-26	To G-9-76
8771	IF-27	To G-9-77
8772	IF-28	To G-9-78
8773	IF-29	
8774	IF-30	
8775	IF-31	To G-9-120
8776	IF-32	To G-9-100
8777	IF-33	
8778	IF-34	To G-9-96
8779	IF-35	
8780	IF-36	To G-9-141
8781	IF-37	To G-9-79
8782	IF-38	
8783	IF-39	
8784	IF-40	
8785	IF-41	To G-9-80
8786	IF-42	
8787	IF-43	To G-9-117
8789	IF-44	To G-9-106
8790	IF-45	
8791	IF-46	
8792	IF-47	
8793	IF-48	To G-9-81
8794	IF-49	To G-9-159
8795	IF-50	To G-9-134
8796	IF-51	To G-9-82
8797	IF-52	
8958	IF-53	To G-9-186
8959	IF-54	To G-9-116
8802	IF-55	
8803	IF-56	To G-9-101
8804	IF-57	
8805	IF-58	
8806	IF-59	To G-9-125
8807	IF-60	To G-9-145
8960	IF-61	
8812	IF-62	
8813	IF-63	
8814	IF-64	To G-9-142
8815	IF-65	
8816	IF-66	To G-9-138
8817	IF-67	
8967	IF-68	To G-9-83
8968	IF-69	To G-9-129
8969	IF-70	To G-9-160
8822	IF-71	To G-9-151
8823	IF-72	To G-9-150
8824	IF-73	
8825	IF-74	To G-9-109
8826	IF-75	To G-9-84
8827	IF-76	
8831	IF-77	To G-9-118
8832	IF-78	To G-9-163
8833	IF-79	To G-9-85
8834	IF-80	To G-9-86
8835	IF-81	
8836	IF-82	To G-9-146
8837	IF-83	
8842	IF-84	To G-9-87
8843	IF-85	To G-9-139
8844	IF-86	To G-9-110
8845	IF-87	To G-9-149
8847	IF-88	To G-9-88
8848	IF-89	To G-9-161
8849	IF-90	
8852	IF-91	To G-9-122
8853	IF-92	
8854	IF-93	To G-9-126
8855	IF-94	To G-9-98
8860	IF-95	
8861	IF-96	To G-9-121
8862	IF-97	To G-9-89
8863	IF-98	To G-9-137
8864	IF-99	To G-9-135
8865	IF-100	
8866	IF-101	To G-9-114
8872	IF-102	
8873	IF-103	
8874	IF-104	To G-9-130
8875	IF-105	
8876	IF-106	To G-9-104
8878	IF-107	To G-9-90
8879	IF-108	To G-9-105
8880	IF-109	
8882	IF-110	To G-9-147
8883	IF-111	
8884	IF-112	To G-9-119
8885	IF-113	To G-9-115
8886	IF-114	To G-9-91
8892	IF-115	To G-9-152
8893	IF-116	To G-9-140
8894	IF-117	To G-9-143
8895	IF-118	
8896	IF-119	
8897	IF-120	To G-9-136
8898	IF-121	
8902	IF-122	To G-9-92
8903	IF-123	To G-9-144
8904	IF-124	To G-9-113
8905	IF-125	
8909	IF-126	To G-9-93
8910	IF-127	To G-9-157
8911	IF-128	To G-9-131
8912	IF-129	To G-9-102
8913	IF-130	
8914	IF-131	To G-9-111
8915	IF-132	To G-9-132
8922	IF-133	
8923	IF-134	
8924	IF-135	To G-9-153
8931	IF-136	
8932	IF-137	To G-9-156
8933	IF-138	To G-9-155
8940	IF-139	
8941	IF-140	To G-9-94
8942	IF-141	To G-9-103
8949	IF-142	To G-9-95
8950	IF-143	To G-9-99
8951	IF-144	To G-9-112

After remanufacture by Hawkers, the Hunters were sold as follows (listed in order of delivery to U.K.) —

G-9- Serial	Ex- IF-	New mark	Customer & new serial
70	6	F.59	Iraq 570
71	10	F.59	Iraq 584
72	14	F.59	Iraq 572
73	20	F.59	Iraq 627
74	21	F.59	Iraq 574
75	24	F.59	Iraq 577
76	26	F.57A	Kuwait 212
77	27	F.59	Iraq 575
78	28	F.59	Iraq 573
79	37	T.67	Kuwait 211
80	41	F.57A	Kuwait 213
81	48	F.59	Iraq 571
82	51	F.59	Iraq 581
83	68	T.69	Iraq 568
84	75	F.59	Iraq 587
85	79	F.59	Iraq 628
86	80	F.59	Iraq 630
87	84	T.69	Iraq 567
88	88	F.59	Iraq 578
89	97	T.69	Iraq 626
90	107	F.59	Iraq 580
91	114	F.59	Iraq 586
92	122	F.59	Iraq 579
93	126	F.59	Iraq 585
94	140	F.59	Iraq 576
95	142	F.57A	Kuwait 212 ntu
	convt'd	F.59	Iraq '580'
	re-serial	F.59A	Iraq 582

Dassault Mystère 20E CM-01
Two Mystère 20 high speed VIP transports were delivered to 21 Smaldeel in 1973. The unit's blue Sioux head insignia appears behind the cockpit. *(S. G. Richards)*

De Havilland DH.82A Tiger Moth T-24
is preserved in the Musée de l'Armée with the spurious code 'UR-!' and the 'comet' marking of 2 Smaldeel on the engine cowling. *(Tony Lloyd)*

De Havilland DH.98 Mosquito NF.30 MB-11. Mosquito NF.30 all-weather fighters equipped two squadrons at Bevekom. MB-11 served with 11 Smaldeel and was sold for scrap in 1956. *(via Flash)*

De Havilland DH.82A Tiger Moth T-7 began its Belgian career with the training school at RAF Snailwell in February 1945 as DF179, later becoming ETA-7 and T-7 with the EVO. *(via F. Binnemans)*

Dornier Do 27D-J1 D04 is a relatively recent addition to the Musée de l'Armée collection, ex 16 Smaldeel. It stands beside a Super Cub bearing the 'OL-' (Observation Leger) prefix which was dropped by the Landmacht during 1974. *(Tony Lloyd)*

De Havilland DH.98 Mosquito NF.30 MB-24 was the last of the night fighter variants delivered to 1 Wing in late 1951, passing to the Musée de l'Armée in March 1957. *(Tony Lloyd)*

Dornier Do 27J-1 OL-DO8 seen on a visit to Middle Wallop in July 1971. Twelve Do 27 were used for liaison duties by 15 and 16 Smaldelen of the Landmacht until replacement by Islanders in 1976. *(S.G. Richards)*

Douglas C-47 Dakota K-10 photographed during a visit to Blackbushe in September 1958 on VIP transport duties; note the large style presentation of the serial and coding compared with the later version below. K-10 was flown to Koksijde for retirement on 26 January 1976. *(MAP)*

Douglas C-47 Dakota K-8 was delivered to Belgium in October 1946, and after service in the Congo and at home was officially withdrawn in January 1976, though it in fact continued to fly, as evidenced by this shot at Melsbroek in August 1976. *(Tony Lloyd)*

Douglas DC-4 KX-1
21 Smaldeel operated two DC-4 aircraft on communications duties to the Congo. KX-1, seen visiting Heathrow in July 1959, eventually became a cafe at Torhout. *(MAP)*

Douglas DC-6AC KY-4 was previously a Sabena aircraft, and with three others formed the long range transport element of 21 Smaldeel. Seen here on a visit to the U.S.A., at Andrews AFB. *(Scotpic)*

Fairchild C-119G Flying Boxcar CP-16 otherwise OT-CAP, with the green dorsal fin-stripe markings of 40 Smaldeel visiting Birmingham. The Flying Boxcars formed the main transport element of the Belgian Air Force from 1951 to 1973. *(N.P. Lewis)*

Fairchild C-119G Flying Boxcar CP-18 OT-CAR seen landing at Brustem in 1968; an unusual shot showing the large call sign presentation. *(P. Glas)*

Fouga CM.170R Magister MT-24 currently serves with the 7 and 9 Smaldelen pool at Brustem, and illustrates the natural metal and dayglo colours applied to the aircraft not used for aerobatics. *(via A.D. Annis)*

FBN BN.2A-21 Islander B-02 carries the last two of its call sign OTALA on the nose. Twelve of the type replaced Dornier 27s with 16 Smaldeel and the SLV during 1976-7. *(J.-L. Gaynecoetche)*

Fouga CM.170R Magister MT-18 appears in the red colours with golden lion insignia of the internationally known 'Rode Duivels/Diables Rouges' aerobatic team. The team's Magisters are normally based at Brustem with the EPE/EVS. *(S.G. Richards)*

18
275

96	34	T.69	private venture
	convt'd	T.66C	Lebanon L282
97	11	F.59	Iraq 631
98	94	F.59	Iraq 629
99	143	T.69	Iraq 593
100	32	F.59	Iraq 583
101	56	T.67	Kuwait 210
102	129	F.70	Lebanon L178
103	141	F.71	Chile J700
104	106	F.71	Chile J701
105	108	F.71	Chile J702
106	44	F.71	Chile J703
107	2	F.56A	India A484
108	22	F.59A	Iraq 659
109	74	F.59A	Iraq 673
	reserialled		Iraq 695
110	86	F.70A	Lebanon L176
111	131	F.56A	India A485
112	144	F.56A	India A486
113	124	F.56A	India A480
114	101	F.70	Lebanon L179
115	113	F.56A	India A487
116	54	F.59A	Iraq 670
	reserialled		Iraq 692
117	43	F.56A	India A488
118	77	F.56A	India A461
119	112	T.66C	Lebanon L281
120	31	F.59A	Iraq 674
	reserialled		Iraq 696
121	96	F.70	Lebanon L177
122	91	F.56A	India A459
123	16	F.56A	India A462
124	4	F.56A	India A460
125	59	F.59A	Iraq 675
	reserialled		Iraq 697
126	93	F.59A	Iraq 676
	reserialled		Iraq 698
127	8	F.59A	Iraq 660
128	13	F.56A	India A463
129	69	F.57A	Kuwait 214
130	104	F.56A	India A464
131	128	F.56A	India A484
132	132	F.56A	India A490
133	9	F.59A	Iraq 661
134	50	F.56A	India A465
135	99	F.59A	Iraq 677
	reserialled		Iraq 699
136	120	F.56A	India A466
137	98	F.56A	India A490
138	66	F.56A	India A491
139	85	F.56A	India A467
140	116	F.56A	India A468
141	36	F.56A	India A492
142	64	F.56A	India A479
143	117	F.56A	India A494
144	123	F.56A	India A470
145	60	T.66C	Lebanon L280
146	82	F.56A	India A469
147	110	F.56A	India A471
148	18	F.56A	India A472
149	87	F.59A	Iraq 669
	reserialled		Iraq 691
150	72	F.59A	Iraq 678
	reserialled		Iraq 700
151	71	F.59A	Iraq 668
	reserialled		Iraq 690
152	115	F.56A	India A473
153	135	F.59A	Iraq 672
	reserialled		Iraq 694
154	7	F.56A	India A474
155	138	F.59A	Iraq 679
	reserialled		Iraq 701
156	137	F.56A	India A481
157	127	F.56A	India A475
158	17	F.56A	India A476
159	49	F.59A	Iraq 680
	reserialled		Iraq 702
160	70	F.57A	Kuwait 215
161	89	F.56A	India A477
162	25	F.59A	Iraq 671
	reserialled		Iraq 693
163	78	F.56A	India A478
186	53	Used for HSA training	
187	1	F.56A	India A482
189	—	Not delivered to HSA	
190	—	Not delivered to HSA	

HAWKER-SIDDELEY HS.748-2A

The Avro 748 short-range twin turboprop transport first flew in June 1960, since when over 300 have been built for world-wide service.
Three aircraft ordered 1974 to complement the Hercules of 15 Wing in the transport role, being equipped with a large freight door. All three were delivered to 21 Smaldeel during the second half of 1976.

C/n	Serial	Remarks	Callsign
1741	CS-01	D/d 28.6.76	BAF41
1742	CS-02	D/d 31.7.76	BAF42
1743	CS-03	Ex G-BEEM	BAF43

LOCKHEED T-33A

Over 6,500 T-33s, a development of the F-80 Shooting Star, were built, and many issued to countries friendly to the U.S.A. The last few years have seen a reduction in European 'T-Birds', but the type will continue in use until the end of the decade, principally in the instrument training role.
Thirty eight aircraft delivered from 10 March 1952 (12 in 1952, 10 in 1953, 2 in 1954, 6 in 1955 and 8 in 1956), and used by Jachtschool, 31 Smaldeel, and 11 Smaldeel/VZZ. Six were loaned to Holland in 1962 to assist in the joint training programme, and followed by two more in 1965-66; the seven survivors of these returned in 1972. Twelve VZZ aircraft, now camouflaged, are due to be replaced by Alpha Jets in 1979.

C/n	Serial	Ex-USAF	Unit	Remarks
5335	FT-1	51-4041		
5337	FT-2	51-4043	11Sm	
5356	FT-3	51-4062	11Sm	Dutch M-59
5446	FT-4	51-4152		Dutch M-43 w/o 27.4.77
5445	FT-5	51-4151		Dutch M-42
5525	FT-6	51-4231	11Sm	Dutch M-44
5527	FT-7	51-4233		Dutch M-45
5994	FT-8	51-6662		Dutch M-46 w/o 8.6.65
5995	FT-9	51-6663		Dutch M-60
5996	FT-10	51-6664		
5993	FT-11	51-6661	11Sm	Dutch M-47
5383	FT-12	51-4089		w/o 15.7.57
6816	FT-13	51-9032		
7032	FT-14	51-9248	11Sm	
7138	FT-15	51-17445	11Sm	
7254	FT-16	51-17463		
7362	FT-17	51-17468	11Sm	
7363	FT-18	51-17469		w/o 28.9.59
7485	FT-19	51-17505		w/o 12.12.61
7373	FT-20	51-17479		w/o 12.1.62
7573	FT-21	51-17513	11Sm	
7584	FT-22	51-17524	11Sm	
7684	FT-23	51-17539		w/o 11.7.61
7788	FT-24	52-9892		
9063	FT-25	53-5724	11Sm	w/o 18.2.76
9064	FT-26	53-5725		w/o 27.4.77
9065	FT-27	53-5726		w/o 13.12.61
9091	FT-28	53-5752	11Sm	
9092	FT-29	53-5753		
9093	FT-30	53-5754	11Sm	
9579	FT-31	55-3038		w/o
9581	FT-32	55-3040		w/o 28.5.57
9582	FT-33	55-3041	11Sm	w/o 1976
9584	FT-34	55-3043		
9587	FT-35	55-3046		w/o 28.9.59
9588	FT-36	55-3047		
9623	FT-37	55-3082	11Sm	
9585	FT-38	55-3044		
	FT-101		11Sm	
	FT-106		11Sm	

In addition, FT-1, -5, -7, -9, -10, -13, -16, -24, -29, -34, -36 and -38 were on charge in mid-1977, presumably in storage. 'FT-101' (noted 3.76 and 12.76) and 'FT-106' (noted 3.76) continue to defy explanation.

Lockheed RT-33A

One example only of this reconnaissance version, withdrawn to storage in the USA in 1968, after use by 9 Wing.

8672 FRT-1 53-5333

LOCKHEED C-130H HERCULES

The famous Hercules transport first flew in 1954, and remains in production, over 1500 having been built for the USAF and military and civil operators world-wide.
In March 1971 twelve aircraft were ordered to replace the C-119G with 15 Wing. The first was delivered on 25 July 1972, and in common with the later deliveries, was flown to Marshalls at Cambridge for finishing, including the application of camouflage. The final aircraft returned from Cambridge to 15 Wing on 16 May 1973. All serve with 20 Smaldeel, and use call-signs 'Belgian Air Force 01 to 12' respectively.

C/n	Serial	USAF Serial
4455	CH-01	71-1797
4460	CH-02	71-1798
4461	CH-03	71-1799
4467	CH-04	71-1800
4470	CH-05	71-1801

4473	CH-06	71-1802
4474	CH-07	71-1803
4478	CH-08	71-1804
4479	CH-09	71-1805
4481	CH-10	71-1806
4482	CH-11	71-1807
4483	CH-12	71-1808

LOCKHEED F-104G STARFIGHTER

Widely used in Europe, the F-104G Starfighter is a multi-role version of the F-104A interceptor first flown on 7 February 1954 and ordered for the USAF in both single and twin-seat configurations. Large-scale international co-operation resulted in production lines being set up in Germany (Messerschmitt), the Netherlands (Fokker), Italy (Fiat) and Belgium (SABCA); the latter three firms also produced aircraft for the Luftwaffe, as well as for the air forces of their respective countries. In Canada, production was undertaken for the RCAF, as well as for several smaller European nations such as Denmark, Greece, Spain and Turkey. Lockheed were responsible for the supply of most of the twin-seat TF-104G trainers. The Starfighter continues to play a principal role in European defence, but is scheduled for replacement by the Tornado and F-16 in the early eighties.

One hundred and thirteen Belgian aircraft (including one replacement) of F-104G and TF-104G marks, issued to 1 Wing (349 and 350 Smaldelen) and 10 Wing (23 and 31 Smaldelen). In addition a conversion flight equipped with three TF-104G was established at Bevekom.

F-104G Starfighter

A total of 100 aircraft ordered from SABCA, increased by one to replace FX-27 which crashed while on test. The remainder of the batch c/ns 9001-9189 were delivered to Germany. The first thirteen aircraft were initially marked FX-01 to FX-013, the first being delivered on 14 February 1963. Deliveries were completed in mid-1965.

C/n	Serial	Wing	Remarks
9016	FX-1	1W	w/o 26.1.71
9017	FX-2	10W	ex1W
9018	FX-3	1W	ex10W
9019	FX-4	10W	ex1W Burnt out 6.76
9020	FX-5	1W	w/o 20.6.68
9021	FX-6	10W	ex 1W w/o 6.4.77
9022	FX-7	10W	ex 1W
9023	FX-8		w/o 16.7.63
9024	FX-9	10W	ex 10W, 1W w/o 3.8.76
9027	FX-10	10W	ex 10W, 1W
9028	FX-11	10W	ex 1W
9029	FX-12	1W	ex 10W
9032	FX-13	1W	ex 1W, 10W
9033	FX-14	1W	w/o 20.7.70
9034	FX-15	1W	
9038	FX-16	1W	w/o 14.12.66
9039	FX-17	10W	ex 1W
9040	FX-18	10W	ex 1W
9044	FX-19	1W	ex 10W
9045	FX-20	1W	ex 1W, 10W
9046	FX-21	10W	ex 1W
9050	FX-22	10W	ex 1W
9051	FX-23	1W	ex 1W, 10W
9052	FX-24	10W	ex 1W
9056	FX-25	1W	ex 1W, 10W w/o 13.1.77
9057	FX-26	10W	ex 1W
9058	FX-27 (1) -	–	w/o 21.11.63
9082	FX-27 (2) -	10W	ex 1W
9062	FX-28	10W	ex 1W
9063	FX-29	10W	
9064	FX-30	10W	ex 1W
9068	FX-31	10W	ex 1W
9069	FX-32	10W	ex 1W
9070	FX-33	10W	
9071	FX-34	10W	ex 1W
9072	FX-35	10W	ex 1W
9073	FX-36	10W	w/o 20.6.68
9077	FX-37	10W	ex 1W w/o 13.11.73
9078	FX-38	10W	
9079	FX-39	1W	
9083	FX-40	10W	
9084	FX-41	10W	
9085	FX-42	10W	w/o 30.9.70
9086	FX-43	10W	w/o 26.1.73
9087	FX-44	1W	
9088	FX-45	1W	ex 10W
9089	FX-46	10W	ex 1W/Slivers w/o 15.1.76
9090	FX-47	10W	
9091	FX-48	10W	ex 1W
9092	FX-49	1W	ex 10W w/o 24.11.69
9093	FX-50		w/o 10.7.72
9094	FX-51	10W	ex 10W, 1W
9095	FX-52	10W	ex 10W, 1W
9096	FX-53	1W	
9097	FX-54	1W	
9098	FX-55		w/o 6.6.69
9099	FX-56	1W	w/o 14.2.77
9100	FX-57	1W	ex 10W
9101	FX-58	1W	ex 10W
9102	FX-59	1W	
9103	FX-60	10W	ex 1W
9104	FX-61	10W	ex 10W, 1W
9105	FX-62	1W	ex 10W
9106	FX-63		w/o 11.10.66
9107	FX-64	10W	ex 1W
9108	FX-65	10W	
9109	FX-66		w/o 18.8.64
9113	FX-67	10W	ex 10W, 1W
9114	FX-68	1W	
9115	FX-69	10W	ex 1W, 10W, 1W
9119	FX-70	10W	
9120	FX-71	10W	w/o 24.9.69
9121	FX-72	1W	ex 10W, 1W, 10W
9125	FX-73	10W	w/o 10.7.72
9126	FX-74	1W	
9127	FX-75	10W	w/o 4.4.68
9131	FX-76	10W	ex 1W
9132	FX-77		w/o 27.1.65
9133	FX-78	1W	
9137	FX-79	10W	
9138	FX-80	1W	ex 10W
9139	FX-81	10W	ex 10W, 1W
9140	FX-82	10W	ex 1W
9141	FX-83	1W	
9142	FX-84	10W	ex 10W, 1W
9146	FX-85	10W	ex 10W, 1W
9147	FX-86	1W	ex 10W
9148	FX-87	10W	w/o 2.9.68
9152	FX-88		w/o 20.4.67
9153	FX-89	10W	ex 1W
9154	FX-90	1W	ex 10W
9158	FX-91	10W	ex 10W, 1W
9159	FX-92		w/o 9.9.65
9160	FX-93	1W	
9164	FX-94	10W	ex 1W, 10W, 1W
9165	FX-95	10W	ex 1W
9166	FX-96	10W	
9170	FX-97	10W	ex 1W w/o 25.7.73
9171	FX-98	10W	
9172	FX-99	10W	ex 10W, 1W
9176	FX-100	10W	ex 1W

Some aircraft reportedly written off have been noted after their supposed accidents, e.g. FX-49 which crashed 24.11.69. In this connection it should be noted that in 1975, at least FX-3 and FX-69 were apparently wooden decoys, although they may still have airworthy 'doubles.'

TF-104G Starfighter

Twelve aircraft delivered. An OCU flight of six aircraft and a simulator was attached to 1 Wing at Bevekom, but current aircraft dispositions indicate that this has moved to 10 Wing at Kleine Brogel.

C/n	Serial	Unit	Remarks
5786	FC-01	10W	ex 1W. Ex 64-15104
5787	FC-02	10W	ex 64-15105
5788	FC-03	10W	ex 1W. Ex 64-15106
5101	FC-04	10W	
5102	FC-05	1W	
5103	FC-06	1W	
5104	FC-07	1W	Believed w/o
5105	FC-08	10W	ex 1W
5106	FC-09		w/o 19.6.69
5107	FC-10	10W	ex 1W
5108	FC-11	10W	
5109	FC-12	10W	ex 1W

During the making of a film FC-08 was reserialled FX-87, and FC-05 became FC-09.

MILES M.14A MAGISTER

A 1937 development of the Miles M.2X and M.2Y Hawk Trainer, the M.14A received the name Magister in military guise, and complemented the Tiger Moth in wartime training and communications work.

One aircraft, ex RAF T9800 c/n 1992 was delivered to Saffraanberg in Oct. 46 for ground instructional use, but was restored to airworthiness, serialled G-1, and was used by 169 Wing 1946-48. It was sold as OO-NIC, but is currently in store for the Luchtmacht museum.

MILES M.25 MARTINET

Developed from the Master, using many common components, the Martinet first flew in 1942 as a specialist target-towing aircraft. 1724 were built, including a handful of remote-control drones, and the type remained in RAF service until replaced by the Tempest TT.5 in the late fifties. Small numbers of ex-RAF aircraft were supplied to Eire, France, Portugal, and Sweden. Belgium also received eleven ex-RAF aircraft for target-towing at Koksijde; all were withdrawn in 1953 when replaced by Meteors.

Serials were R-1 to R-11, but the RAF serial tie-ups were unknown.

First batch of nine delivered 22 March to 16 November 1947, including EG683, HP415 (became R-5), MS773, MS815, MS856, NR297, NR422 and NR441.
Two replacements delivered November 1948, were ex JN539 and NR650. These replaced HP415 and MS773 which returned to the RAF 4 August 1948.

NORTH AMERICAN HARVARD AND TEXAN

Used in large numbers by British and American training units during World War II, the Harvard and Texan were issued to Allied air forces in 1946-47.
Belgium received 213 aircraft, including, officially, 58 transferred from RAF stocks; the balance were provided from re-manufactured USAF aircraft in 1951.

First batch of 52 ex-RAF Harvards delivered 27 January to 13 May 1947 –

Mark IIA – EX181 EX230 EX251
EX254 EX264 EX273 EX275 EX292
EX295 EX303 EX305 EX318 EX371
EX438 EX439 EX448 EX461 EX476
EX542 EX544 EX546 EX547 EX550
EX551 EX567 EX602 EX623 EX633
EX660 EX661 EX680 EX760 EX779
EX821 EX823
Mark III – EX910 EX937 EX939
EX940 EX946 EX958 EX974 EX993
EX944
Mark IIA– EZ162 EZ174 EZ186
Mark III – EZ210 EZ214 EZ256
Mark IIA– EZ292 EZ310

Second batch of 9 ex-RAF Harvards delivered 7 to 29 September 1949 –

Mark IIB – FX212 FX299 FX414
FX466 KF415 KF463 KF483 KF490
KF568

Ex-USAF Texans, e.g. 51-17215/H-144 from batch 51-17089/17231 (type T-6H and T-6J). Possibly all 142 aircraft to Belgium.

Ten Harvards transferred from Holland October 1952 to April 1953 –

B-20/FT142	B-37/FT410
B-47/FT430	B-58/FT286
B-89/FS885	B-99/FT247
B-104/FS728	B-128/FS730
B-131/FT381	B-139/FT390

Eight Harvards loaned by Holland from 21 August 1948 to 18 August 1949 –

B-57/FT220	B-101/FT407
B-103/FT419	B-105/FT210
B-118/FT427	B-119/FS731
B-120/FS820	B-125/FS833

NORTH AMERICAN F-86F SABRE

According to USAF records, 51-13202, 52-5305, 5362, 5367 and 5402 were allocated to Belgium, possibly for evaluation. The type did not enter squadron service with the BLu, however.

PERCIVAL P.31C PROCTOR IV

Developed directly from the Vega Gull, the Proctor was produced extensively during wartime as communications and radio-training aircraft. The final RAF variant, the mark IV, featured a re-designed and strength ened fuselage, and most were built by F. Hills & Son at Manchester with c/ns prefixed 'H-'. Six surplus RAF aircraft were delivered - first four June 1947, P-5 in October 1947, and P-6 10 March 1948. Used as communications aircraft by 367 Squadron, until withdrawn in 1950.

C/n	Serial	Ex-RAF	Remarks
H721	P-1	NP350	
H589	P-2	NP182	To OO-FEB
H575	P-3	NP168	
H578	P-4	NP171	w/o 10.5.49
H571	P-5	NP164	
H654	P-6	NP270	

PERCIVAL PEMBROKE C.51

Belgium was the first overseas customer for this development of the Prince, with twelve aircraft delivered between 18 February and 10 April 1954, all for use by 21 Smaldeel. Some aircraft were fitted with a glazed nose and used by ESCF for photography and navigation training. As the aircraft became due for re-sparring, they were withdrawn, the last four being RM-3, 4, 10, and 11, all withdrawn in 1976, and replaced by the Swearingen Merlin. RM-1, 2, 5, 6, 8, and 12 were all in storage at Koksijde in 1975, whilst RM-7 has been preserved at Molenbeek.

C/n	Serial	Callsigns Old	New
14	RM-1	OTZAA	BAF81
17	RM-2	OTZAB	BAF82
20	RM-3	OTZAC	BAF83
21	RM-4	OTZAD	BAF84
24	RM-5	OTZAE	BAF85
25	RM-6	OTZAF	BAF86
27	RM-7	OTZAG	BAF87
28	RM-8	OTZAH	BAF88
29	RM-9	OTZAI	BAF89
31	RM-10	OTZAJ	BAF90
32	RM-11	OTZAK	BAF91
33	RM-12	OTZAL	BAF92

PIPER L-18C AND L-21B SUPER CUB

The Super Cub was a 1950 development of the well known L-4 Cub (Grasshopper) and was produced widely for civil and military use. MDAP Super Cubs served with the armies of Belgium, Denmark, France, Germany, Greece, Holland, Italy, Norway and Turkey, in two versions, the L-18 with 90 h.p. Continental engine, and the L-21 powered by a 135 h.p. Lycoming.
One hundred and fifty seven aircraft delivered under the Mutual Defense Aid Program, replacing the Auster AOP.6 with the Landmacht, and including 16 passed on to Denmark (OL-L88 to 103) and 9 to Holland (OL-L104 to 112). During 1975 a further five L-21B were purchased from Holland for use as glider tugs for the Luchtcadetten (Air Cadets), though maintained by the air force. Many European Super Cubs have been sold on the civil markets.

L-18C

C/n	Serial	Ex-	Remarks/To
18-1574	OL-L01	51-15574	
18-1575	OL-L02	51-15575	
18-1572	OL-L03	51-15572	
18-1573	OL-L04	51-15573	OO-HSC OO-KIW
18-1576	OL-L05	51-15576	D-EDCI
18-3080	OL-L06	53-4680	
18-3081	OL-L07	53-4681	
18-3082	OL-L08	53-4682	
18-3083	OL-L09	53-4683	D-ELQY
18-3084	OL-L10	53-4684	D-ECBE D-EFOC
18-3085	OL-L11	53-4685	D-EAEB
18-3086	OL-L12	53-4686	

18-3087 OL-L13 53-4687 OO-FER
18-3088 OL-L14 53-4688 LN-LJJ
18-3089 OL-L15 53-4689
18-3090 OL-L16 53-4690
18-3091 OL-L17 53-4691
18-3092 OL-L18 53-4692
18-3093 OL-L19 53-4693
18-3094 OL-L20 53-4694
18-3095 OL-L21 53-4695
18-3096 OL-L22 53-4696
18-3097 OL-L23 53-4697 D-EHCC
18-3098 OL-L24 53-4698
18-3099 OL-L25 53-4699 D-EAUB
18-3100 OL-L26 53-4700
18-3101 OL-L27 53-4701
18-3102 OL-L28 53-4702 D-EDOQ
18-3103 OL-L29 53-4703 LN-LJK
18-3104 OL-L30 53-4704
18-3105 OL-L31 53-4705 LN-LJL
18-3106 OL-L32 53-4706 OO-GDE
18-3107 OL-L33 53-4707
18-3108 OL-L34 53-4708
18-3109 OL-L35 53-4709 LN-BEE
18-3110 OL-L36 53-4710
18-3111 OL-L37 53-4711 PH-WAM
18-3112 OL-L38 53-4712 OO-DPE
18-3113 OL-L39 53-4713 D-ENKC
18-3114 OL-L40 53-4714 D-ELCU
18-3115 OL-L41 53-4715
18-3116 OL-L42 53-4716 D-ECBA
18-3117 OL-L43 53-4717
18-3118 OL-L44 53-4718 LN-LJI
18-3119 OL-L45 53-4719
18-3120 OL-L46 53-4720 D-EOAB
18-3121 OL-L47 53-4721
18-3122 OL-L48 53-4722 LN-LJM
18-3123 OL-L49 53-4723 OO-GDG
18-3124 OL-L50 53-4724
18-3125 OL-L51 53-4725 D-EETU
18-3126 OL-L52 53-4726 D-EFBU
18-3127 OL-L53 53-4727
18-3128 OL-L54 53-4728
18-3129 OL-L55 53-4729 D-EHTR
18-3130 OL-L56 53-4730
18-3131 OL-L57 53-4731 OO-GDH
18-3132 OL-L58 53-4732 OO-LJH
18-3133 OL-L59 53-4733 OO-LJG
18-3134 OL-L60 53-4734
18-3135 OL-L61 53-4735
18-3136 OL-L62 53-4736
18-3137 OL-L63 53-4737 D-EMCD
D-EHCD
18-3138 OL-L64 53-4738

18-3139 OL-L65 53-4739 OO-LJN
18-3140 OL-L66 53-4740 OO-MVZ
18-3141 OL-L67 53-4741 D-EKQG
18-3142 OL-L68 53-4742
18-3143 OL-L69 53-4743
18-3144 OL-L70 53-4744
18-3145 OL-L71 53-4745 D-EAUS
18-3146 OL-L72 53-4746 OO-ACK
18-3147 OL-L73 53-4747
18-3152 OL-L74 53-4752 LX-UXP
18-3153 OL-L75 53-4753
18-3154 OL-L76 53-4754
18-3155 OL-L77 53-4755 D-EKKK
18-3156 OL-L78 53-4756 D-ENOS
18-3157 OL-L79 53-4757
18-3158 OL-L80 53-4758
18-3159 OL-L81 53-4759
18-3160 OL-L82 53-4760
18-3161 OL-L83 53-4761 D-EBFC
18-3162 OL-L84 53-4762 OO-FBA
18-3163 OL-L85 53-4763
18-3148 OL-L86 53-4748
18-3149 OL-L87 53-4749
18-3150 OL-L88 53-4750 Y-651
18-3151 OL-L89 53-4751 Y-652
18-3164 OL-L90 53-4764 Y-653
18-3165 OL-L91 53-4765 Y-654
18-3166 OL-L92 53-4766 Y-655
18-3167 OL-L93 53-4767 Y-656
18-3168 OL-L94 53-4768 Y-657
18-3169 OL-L95 53-4769 Y-658
18-3170 OL-L96 53-4770 Y-659
18-3171 OL-L97 53-4771 Y-660
18-3172 OL-L98 53-4772 Y-661
18-3173 OL-L99 53-4773 Y-662
18-3174 OL-L100 53-4774 Y-663
18-3175 OL-L101 53-4775 Y-664
18-3176 OL-L102 53-4776 Y-665
18-3177 OL-L103 53-4777 Y-666
18-3178 OL-L104 53-4778 R-80
18-3179 OL-L105 53-4779 R-81
18-3180 OL-L106 53-4780 R-82
18-3181 OL-L107 53-4781 R-83
18-3182 OL-L108 53-4782 R-84
18-3183 OL-L109 53-4783 R-85
18-3184 OL-L110 53-4784 R-86
18-3185 OL-L111 53-4785 R-87
18-3186 OL-L112 53-4786 R-88
18-3187 OL-L113 53-4787 D-EHTV
18-3188 OL-L114 53-4788 OO-WIK
18-3189 OL-L115 53-4789 D-EHCH
18-3190 OL-L116 53-4790 D-EHCF
18-3191 OL-L117 53-4791 OO-HBB

18-3192 OL-L118 53-4792 PH-VCW
18-3193 OL-L119 53-4793 D-ENKB
18-3194 OL-L120 53-4794
18-3195 OL-L121 53-4795 D-EGMO
18-3196 OL-L122 53-4796 D-ECZD
18-3197 OL-L123 53-4797 D-EFBD
18-3198 OL-L124 53-4798 OO-HLT
18-3199 OL-L125 53-4799 D-EABS
18-3200 OL-L126 53-4800 D-ELSG
18-3201 OL-L127 53-4801
18-3202 OL-L128 53-4802 D-ELFT
18-3203 OL-L129 53-4803
18-3204 OL-L130 53-4804 D-ECZF
18-3205 OL-L131 53-4805 OO-LVM
18-3206 OL-L132 53-4806 D-EJQO
18-3207 OL-L133 53-4807
18-3208 OL-L134 53-4808
18-3209 OL-L135 53-4809 D-ELFZ
18-3210 OL-L136 53-4810 OO-LMM
18-3211 OL-L137 53-4811
18-3212 OL-L138 53-4812 OO-DPE
18-3213 OL-L139 53-4813 LX-UXC
18-3214 OL-L140 53-4814 OO-ACF
OO-AAP
18-3215 OL-L141 53-4815
18-3216 OL-L142 53-4816 OO-ACG
18-3217 OL-L143 53-4817 D-EBFD
18-3218 OL-L144 53-4818 OO-LOT
18-3219 OL-L145 53-4819 D-EHCB
18-3220 OL-L146 53-4820 D-EHTS
18-3221 OL-L147 53-4821 OO-ATY
18-3222 OL-L148 53-4822
18-3223 OL-L149 53-4823 OO-HBA
18-3224 OL-L150 53-4824 D-EKOH
18-3225 OL-L151 53-4825
18-3226 OL-L152 53-4826
18-3227 OL-L153 53-4827 OO-HBC
18-3228 OL-L154 53-4828 OO-HBG
18-3229 OL-L155 53-4829 D-EEMS
18-3230 OL-L156 53-4830 OO-ACC
18-3231 OL-L157 53-4831

L-21B

Serial	Remarks
LB-01	
LB-02	
LB-03	
LB-04	
LB-05	ex 54-2464, R-174

One of the others is ex 54-2444 and R-154

N.B. Six aircraft were also delivered to the Belgian Congo under MDAP.

REPUBLIC F-84E/G THUNDERJET

Early models of the F-84 were produced in significant numbers for the USAF, and used operationally in the Korean War. Small numbers of the F-84E were made available to NATO countries in Europe, but with the slow evolution of the swept-wing F-84F variant, the F-84G was introduced as a stopgap. By the time the more advanced F-84F was ready, 3025 F-84G Thunderjets had been built, many of which saw service in Europe.
Two hundred and thirteen aircraft delivered to Belgium, commencing 18 April 1951, and allocated to 2, 9, and 10 Wings. They were replaced by the F-84F in 1956, with the aircraft marked 'D' below being passed to Denmark.

F-84E

Ex-USAF	Serial	Remarks
50-1819	FS-1	
50-1826	FS-2	
50-1825	FS-3	
50-1813	FS-4	
51-9601	FS-5	
51-9607	FS-6	
51-9581	FS-7	
51-9603	FS-8	
51-9605	FS-9	
51-9608	FS-10	
51-9609	FS-11	
51-9559	FS-12	
51-9573	FS-13	
51-9579	FS-14	
51-9602	FS-15	
51-9606	FS-16	
51-9599	FS-17	
51-9613	FS-18	
51-9614	FS-19	
51-9616	FS-20	
51-9615	FS-21	

F-84G

Known aircraft and serial tie-ups only –

Ex-USAF	Serial	Remarks
51-9627		
51-9637		D

51-9671		
51-9673		
51-9674	FZ-11	D
51-9675	FZ-19	D
51-9676	FZ-7	D
51-9678		
51-9681	FZ-10	D
51-9684		
51-9694		
51-9696		
51-9702		
51-9707	FZ-2	D
51-9709	FZ-12	D
51-9711	FZ-15	D
51-9712	FZ-16	D
51-9715		
51-9710		
51-9721		
51-9723	FZ-9	D
51-9736		
51-9749		
51-9750		
51-9762		
51-9764		
51-9770		
51-9772		
51-9774	FZ-24	D
51-9787		
51-9801		
51-9828		
51-9837		
51-9839		
51-9840		
51-9844		D
51-9845		
51-9857		
51-9923		
51-9936		
51-9944	FZ-43	D
51-9949	FZ-57	D
51-9956		
51-9957		
51-9958		D
51-9961		
51-9962		
51-9965		
51-9968		
51-9975		
51-9980		
51-9986	FZ-49	D
51-9992		
51-9997		D
51-9999		
51-10011		
51-10012		
51-10018	FZ-46	D
51-10020		
51-10025		
51-10033		
51-10034		
51-10042		
51-10046		
51-10047		
51-10076	FZ-75	D
51-10180		
51-10195		
51-10199		
51-10200		
51-10223		
51-10233		
51-10236		
51-10297	FZ-86	D
51-10298	FZ-76	D
51-10487	FZ-77	D
51-10501	FZ-82	D
51-10611	FZ-105	D
51-10622	FZ-118	D
51-10629	FZ-92	D
51-10708	FZ-120	D
51-10731	FZ-100	D
51-10744	FZ-102	D
51-10747	FZ-125	D
51-10751	FZ-103	D
51-10798	FZ-141	D
51-10799	FZ-144	D
51-10802	FZ-143	D
51-10820	FZ-154	D
51-10902	FZ-138	D
51-10909	FZ-131	D
51-10917	FZ-136	D
51-10930	FZ-132	D

REPUBLIC F-84F THUNDERSTREAK

Originally conceived as the F-96, the F-84F was a swept-wing version of the Thunderjet, and first flown on 3 June 1950. Difficulties with the wing structure and the US-built model of the Armstrong-Siddeley Sapphire engine delayed introduction of the F-84F until 1955, when the first were supplied to NATO countries in Europe, from a production run which eventually totalled 2,713. One hundred and ninety seven Belgian aircraft used by 2 Wing (1, 2 and 3 Smaldelen) and 10 Wing (23, 27 and 31 Smaldelen). Deliveries of the main batch of 180 began on 4 June 1955, and were completed with FU-178 on 29 January 1958. A further seventeen from surplus USAF stocks were added in April 1958. The type was replaced during the early sixties by the F-104G Starfighter and Mirage 5B. Of the 105 aircraft written off in service, only 30 were damaged beyond economical repair (BER in the lists below) in minor accidents, the rest being totally destroyed in crashes. These high attrition figures are parallel with F-84F experience in other air forces. On retirement, at least 74 of the surviving aircraft were placed in store at Koksijde in 1971, and from here a few have passed on: FU-6 to the Southend Museum 14.11.73; FU-30 to the Brussels Museum; FU-174 to Saffraanberg Technical School with FU-187; and FU-74 and FU-93 to Bierset for decoy duties with wooden nose and tail extensions so that they resemble the Mirage.

Ex-USAF	Serial	Codes	Remarks
52-7116	FU-1	3R-L	w/o 23.5.56
52-7117	FU-2	YL-C	w/o 17.7.63
52-7118	FU-3	YL-J	w/o 8.7.64
52-7123	FU-4	YL-B,YL-M	BER 7.7.70
52-7124	FU-5	3R-B,3R-E	
52-7133	FU-6	YL-A	
52-7125	FU-7	YL-F	BER 15.6.68
52-7141	FU-8	3R-V,3R-P YL-Y	w/o 9.7.63
52-7148	FU-9	YL-G	w/o 8.12.55
52-7115	FU-10	3R-J	
52-7127	FU-11	3R-C	
52-7137	FU-12	3R-D,YL-V	w/o 21.8.67
52-7139	FU-13	YL-P	w/o 28.11.55
52-7144	FU-14	3R-S	w/o 28.7.58
52-7154	FU-15	YL-M,YL-B YL-F	w/o 30.3.56
52-7135	FU-16	YL-D,UR-S	w/o 29.1.71
52-7136	FU-17		
52-7138	FU-18	YL-Q	BER 23.4.56
52-7143	FU-19	UR-I,YL-L	w/o 12.10.61
52-7151	FU-20	3R-C,3R-V	Storage
52-7170	FU-21	YL-C	Decoy
52-7171	FU-22		
52-7152	FU-23	3R-F,3R-E, YL-R,Z6-H	w/o 24.8.59
52-7156	FU-24	YL-J	w/o 17.10.58
52-7161	FU-25	3R-W	w/o 25.10.56
52-7162	FU-26	3R-T	
52-7163	FU-27	3R-B	w/o 19.2.62
52-7166	FU-28	YL-K	Decoy
52-7175	FU-29	UR-D	
52-7169	FU-30	3R-K	Preserved
52-7178	FU-31	3R-C	
52-7183	FU-32	UR-K,YL-K	w/o 13.5.65
52-7192	FU-33	3R-B	
52-7195	FU-34		BER 5.6.64
52-7104	FU-35	3R-K	w/o 11.7.56
52-7157	FU-36	YL-R	
52-7180	FU-37		
52-7182	FU-38	YL-B	w/o 1.10.57
52-7194	FU-39	YL-A	w/o 29.1.71
52-7187	FU-40	YL-I,YL-U	w/o 16.10.59
52-7193	FU-41	3R-Y	w/o 4.8.66
52-7196	FU-42	YL-W,UR-S	w/o 9.7.63
52-7202	FU-43	3R-P	w/o 23.7.58
52-7207	FU-44	3R-M,3R-B	
52-7210	FU-45		Decoy
52-7220	FU-46	3R-L	w/o 19.6.57
52-7222	FU-47	3R-E	BER 18.6.58
52-7223	FU-48	UR-T,UR-L	w/o 17.1.62
52-7008	FU-49	3R-D	Decoy
52-7011	FU-50	3R-T	Decoy
52-7215	FU-51	3R-G	Decoy
52-7216	FU-52	3R-V,3R-W YL-F	
52-7225	FU-53	YL-X	w/o 4.6.58
52-7226	FU-54	YL-B,UR-B	w/o 22.3.61
52-10521	FU-55	YL-E	
52-10526	FU-56	3R-S	w/o 26.3.58
52-10532	FU-57	YL-H	BER 24.4.58
52-10537	FU-58	3R-Z,UR-N	BER 13.2.68
53-6543	FU-59	RA-R	
53-6549	FU-60	Z6-S	w/o 26.1.68
52-7012	FU-61	UR-H	w/o 3.6.64
52-7129	FU-62	3R-Y	w/o 21.5.58
52-10519	FU-63	YL-D,UR-W	Decoy
52-10531	FU-64	3R-I	BER 30.1.64
53-6560	FU-65	RA-S	w/o 18.3.57
53-6677	FU-66	Z6-D	Decoy
53-6681	FU-67	RA-H	Decoy
53-6683	FU-68	Z6-V	w/o 14.12.56
53-6685	FU-69	Z6-X	BER 6.3.57
53-6698	FU-70		w/o 2.7.57
53-6703	FU-71	RA-O	w/o 2.7.57
53-6670	FU-72	RA-F	w/o 12.3.59

53-6773	FU-73		
53-6777	FU-74		Decoy
53-6687	FU-75	UR-Y	w/o 13.2.58
53-6536	FU-76	UR-L	Decoy
53-6538	FU-77	Z6-M	w/o 18.10.58
53-6541	FU-78	UR-K	BER 9.7.62
53-6576	FU-79	UR-Y	w/o 1.10.56
53-6573	FU-80		BER 10.8.65
53-6577	FU-81	UR-E	BER 30.5.60
53-6587	FU-82		Decoy
53-6555	FU-83	RA-K	BER 18.3.57
53-6564	FU-84	RA-K	w/o 6.8.64
53-6571	FU-85	Z6-Y,8S-C	
53-6583	FU-86	RA-P,Z6-V	w/o 26.10.66
53-6589	FU-87	Z6-O,RA-D	w/o 18.5.62
53-6598	FU-88		BER 14.9.58
53-6675	FU-89	RA-Q	w/o 16.4.58
53-6679	FU-90	RA-A	w/o 1.6.60
53-6693	FU-91	Z6-R,8S-R	
53-6702	FU-92	Z6-P,RA-E	
53-6717	FU-93	RA-B	Decoy
53-6602	FU-94	UR-M	w/o 9.11.70
53-6581	FU-95	8S-I	BER 21.4.70
53-6599	FU-96	UR-C,Z6-W	w/o 15.3.62
53-6539	FU-97	8S-Y	
53-6565	FU-98	8S-G	w/o 16.10.62
53-6618	FU-99		BER 13.10.59
53-6562	FU-100	8S-L	w/o 19.1.59
53-6540	FU-101		w/o 16.10.62
53-6561	FU-102	RA-G	w/o 17.2.57
53-6597	FU-103	RA-J,Z6-L	
53-6701	FU-104	RA-Z,RA-G	BER 22.1.63
53-6707	FU-105	RA-C,RA-D	
53-6722	FU-106	8S-I,RA-F	
53-6684	FU-107	RA-B	BER 16.10.58
53-6610	FU-108		BER 15.4.70
53-6593	FU-109	Z6-B	BER 25.9.57
53-6615	FU-110	Z6-G,3R-T UR-N	w/o 17.1.62
53-7199	FU-111	3R-W	w/o 17.10.58
53-6690	FU-112		w/o 12.6.63
53-6700	FU-113	RA-R	
53-6553	FU-114	Z6-L	w/o 12.9.62
53-6609	FU-115	Z6-E	w/o 22.4.58
53-6738	FU-116	RA-S,Z6-E	Decoy
53-6739	FU-117	Z6-S,RA-O	
53-6580	FU-118	Z6-B	
53-6720	FU-119	RA-P,UR-G	
53-6729	FU-120		w/o 23.6.69
53-6753	FU-121	Z6-N	w/o 20.6.66
53-6755	FU-122	RA-L	
53-6768	FU-123	RA-N,Z6-H	
53-6750	FU-124	Z6-O	w/o 20.6.66
53-6760	FU-125	8S-C	
53-6680	FU-126		w/o 7.5.62
53-6746	FU-127	8S-D	w/o 30.9.57
53-6749	FU-128	Z6-A	w/o 20.3.59
53-6751	FU-129	8S-F,8S-H	w/o 28.1.60
53-6769	FU-130	RA-M	BER 21.8.57
53-6761	FU-131	8S-H	BER 12.8.57
53-6601	FU-132	RA-M	w/o 3.2.67
53-6762	FU-133	Z6-D,RA-C	
53-6759	FU-134		
53-6559	FU-135	UR-F	w/o 28.10.61
53-6607	FU-136	Z6-R	w/o 27.1.61
53-6554	FU-137	Z6-A	BER 20.2.69
53-6782	FU-138	8S-C	w/o 9.2.68
53-6781	FU-139	8S-S,8S-L	
53-6551	FU-140		
53-6747	FU-141	8S-N,Z6-J	w/o20.10.63
53-6574	FU-142		
53-7691	FU-143	8S-Y	
53-6783	FU-144	8S-F,RA-I	
53-6613	FU-145	Z6-E	
53-6532	FU-146	8S-B	
53-6793	FU-147	8S-Q,UR-N	BER 22.4.59
53-6546	FU-148		
53-6797	FU-149	Z6-A,8S-V RA-O	
53-6830	FU-150	8S-R	w/o 26.7.67
53-6854	FU-151		
53-6765	FU-152	RA-P	
53-6784	FU-153	8S-M,8S-S	
53-6806	FU-154		
53-6811	FU-155	8S-R	
53-6834	FU-156		
53-6885	FU-157	UR-S	w/o 11.10.57
53-6891	FU-158	3R-P	
53-6824	FU-159	8S-T	
53-6899	FU-160	UR-U,UR-V	BER 5.5.70
53-6906	FU-161	UR-E,UR-Y	w/o 25.3.64
53-6919	FU-162	UR-V	BER 22.6.60
53-6933	FU-163	UR-L,3R-S	
53-6947	FU-164		
53-6951	FU-165	3R-L	w/o 28.4.65
53-6928	FU-166	UR-S,UR-M	
53-6938	FU-167	UR-N,UR-A	
53-6939	FU-168	UR-K	BER 8.5.67
53-6956	FU-169	UR-Y	w/o 23.11.64
53-6792	FU-170	RA-E	w/o 12.3.69
53-6873	FU-171	3R-K	
53-6900	FU-172	3R-H	
53-6913	FU-173		
53-6862	FU-174	UR-G,3R-L	Instructional
53-6864	FU-175	UR-U,UR-C	
53-6914	FU-176	UR-D	w/o 13.5.58
53-6888	FU-177	UR-Z	
53-6929	FU-178	UR-V,UR-U	w/o 12.7.62
53-6941	FU-179	UR-J	
53-6870	FU-180	UR-A	w/o 4.7.60
52-6371	FU-181	8S-D	
52-6375	FU-182		w/o 2.7.62
52-6403	FU-183	Z6-R	
52-6430	FU-184		
52-6569	FU-185	8S-U,Z6-D	
52-6620	FU-186	8S-R,Z6-E	
52-7067	FU-187	Z6-J,3R-Q	Instructional
52-6369	FU-188	Z6-D	
52-6424	FU-189		w/o 26.10.66
52-7034	FU-190	Z6-V	
52-6374	FU-191	8S-K	
52-6605	FU-192	8S-H	
52-7059	FU-193	Z6-U,3R-N	w/o 27.5.60
52-6378	FU-194		
52-6417	FU-195	Z6-X	w/o 5.5.64
52-7069	FU-196	8S-T	BER 8.4.63
52-6584	FU-197	8S-J	

REPUBLIC RF-84F THUNDERFLASH

A photo-reconnaissance version of the F-84F fitted with extended nose and bifurcated air intakes, produced between March 1954 and January 1958.
The Luchtmacht received an initial batch of 25, commencing 8 June 1955. These were augmented by five more in 1963, and finally by four ex-Luftwaffe examples in 1965. All were used by 42 Verkennings-smaldeel, until the type was replaced by the Mirage 5BR in 1971 and the surviving aircraft ferried to Koksijde, whence three have escaped for preservation: FR-27 at Bevekom; FR-28 at Brussels museum, and FR-34 at Bierset.

Ex-USAF	Serial	Codes	Remarks
51-11250	FR-1	H8-A	w/o
51-16696	FR-2	H8-B	w/o
51-16998	FR-3	H8-C	
52-7240	FR-4	H8-D	
51-1949	FR-5	H8-E	
51-1958	FR-6	H8-F	
51-1957	FR-7	H8-G	w/o
51-11257	FR-8	H8-H	
52-7229	FR-9	H8-J	
52-7230	FR-10	H8-K	w/o
51-16999	FR-11	H8-L	w/o
52-7238	FR-12	H8-M	
52-7237	FR-13	H8-N	w/o
52-7281	FR-14	H8-O	w/o
52-7282	FR-15	H8-P	w/o
52-7279	FR-16	H8-Q	w/o
52-7297	FR-17	H8-R	
51-11297	FR-18	H8-S	
52-7439	FR-19	H8-T	
52-7436	FR-20	H8-U	w/o
51-1876	FR-21	H8-V	
51-1906	FR-22	H8-W	
51-1912	FR-23	H8-X	
51-1898	FR-24	H8-Y	w/o
51-1867	FR-25	H8-	w/o 19.6.69
51-1886	FR-26	H8-	w/o 23.10.70
51-1922	FR-27	H8-	
51-1945	FR-28	H8-	
51-11279	FR-29	H8-	
51-17015	FR-30	H8-	
53-7644	FR-31		ex EA311
53-7646	FR-32		ex EA303
53-7658	FR-33		ex EA305
53-7677	FR-34		ex EA334

SIAI SF.260MB

Developed from the prototype Aviamilano F.250 flown on 15 July 1964, the SF.260 with its high power-to-weight ratio has been ordered by several air forces throughout the world for training (including full aerobatics), and counter-insurgency duties. Thirty-six aircraft ordered January 1969 as replacements for the Stampe SV.4s at EVS, Goetsenhoven, with deliveries between 27 November 1969 and Spring 1971.

C/n	Serial	
10-01	ST-01	w/o 12.4.71
10-02	ST-02	
10-03	ST-03	
10-04	ST-04	
10-05	ST-05	
10-06	ST-06	
10-07	ST-07	

10-08	ST-08	
10-09	ST-09	
10-10	ST-10	w/o
10-11	ST 11	
10-12	ST-12	
10-13	ST-13	w/o 7.72 and converted to simulator
10-14	ST-14	
10-15	ST-15	
10-16	ST-16	
10-17	ST-17	
10-18	ST-18	
10-19	ST-19	
10-20	ST-20	
10-21	ST-21	
10-22	ST-22	
10-23	ST-23	
10-24	ST-24	
10-25	ST-25	
10-26	ST-26	
10-27	ST-27	
10-28	ST-28	
10-29	ST-29	
10-30	ST-30	
10-31	ST-31	
10-32	ST-32	
10-33	ST-33	
10-34	ST-34	
10-35	ST-35	
10-36	ST-36	

SIKORSKY S-55

Sikorsky's ubiquitous utility helicopter was produced in large numbers during the fifties, and saw world-wide service in both military and civil guises.
At least two aircraft were used in the Belgian Congo, both ex Sabena. S-42 was taken over by the Katangans during the 1960 civil war.

Ex-	Serial	Remarks
OO-CWE	S-41	
OO-CWF	S-42	To KAT-42

SIKORSKY-SUD S-58

Produced from 1954 as the H-34 Choctaw for the US Army, and the HSS-1 Seabat for the US Navy, the S-58 found its way into service in several European countries under MDAP, and through licence production in France by Sud Aviation, and in the U.K. by Westland. Production ceased in 1971, and many military examples have been civilianised, and converted to turbine power. Five HSS-1 delivered in May 1961 by Sud Aviation from their licence production batch, to the SAR Flight at Koksijde, with serials following on from those of the Bristol Sycamores. These later augmented by seven ex-Sabena Sikorsky S-58 models, two of which, B-7 and B-8 were on permanent detachment to the Zeemacht. The type was formally replaced by the Westland Sea King in 1976, but B-4, 5 and 6 were all retained in service during 1977.

C/n	Serial	Callsign	Ex-	Remarks
SA145	B-4	OTZKD		40Sm
SA146	B-5	OTZKE		40Sm
SA181	B-6	OTZKF		40Sm
SA184	B-7	OTZKG		w/o 7.1.76
SA185	B-8	OTZKH		40Sm
58-324	B-9	OTZKI	OO-SHG	40Sm
58-333	B-10	OTZKJ	OO-SHH	40Sm
58-356	B-11	OTZKK	OO-SHI	40Sm
58-388	B-12	OTZKL	OO-SHL	40Sm
58-395	B-13	OTZKM	OO-SHM	Brussels Museum
58-850	B-14	OTZKN	OO-SHQ	40Sm
58-826	B-15	OTZKP	OO-SHP	40Sm

STAMPE SV-4B/C

Outwardly resembling its contemporary, the Tiger Moth, the SV-4 was returned to production after the war for the air arms of Belgium and France. Two versions were involved - the SV-4B built by Societe Stampe et Renard in Belgium, with a Gipsy Major 1 engine, and the SV-4C built by SNCA du Nord in France, powered by the Renault 4P. A total of sixty five aircraft replacing the Tiger Moths with the EVS at Goetsenhoven, in two batches: 20 SV-4Bs between 29 June 1948 and 3 December 1948, and 45 SV-4C delivered between 12 March 1951 and 7 October 1955. The type was withdrawn 1970-71 in favour of the SIAI SF.260MB.

C/n	Serial	Remarks
1143	V-1	
1144	V-2	
1145	V-3	
1146	V-4	
1147	V-5	To OO-PAX
1148	V-6	
1149	V-7	
1150	V-8	
1151	V-9	
1152	V-10	
1153	V-11	
1154	V-12	
1155	V-13	
1156	V-14	
1157	V-15	
1158	V-16	
1159	V-17	
1160	V-18	
1161	V-19	
1162	V-20	
1163	V-21	Preserved at Goetsenhoven
1164	V-22	
1165	V-23	
1166	V-24	
1167	V-25	
1168	V-26	
1169	V-27	To G-AZUL
1170	V-28	To OO-CLH
1171	V-29	To OO-GWB
1172	V-30	
1173	V-31	
1174	V-32	
1175	V-33	
1176	V-34	
1177	V-35	
1178	V-36	
1179	V-37	
1180	V-38	
1181	V-39	
1182	V-40	
1183	V-41	To OO-LYK
1184	V-42	
1185	V-43	
1186	V-44	
1187	V-45	
1188	V-46	Preserved at Brussels
1189	V-47	To OO-MCI
1190	V-48	To OO-PAM
1191	V-49	
1192	V-50	
1193	V-51	
1194	V-52	Preserved at Brustem
1195	V-53	
1196	V-54	To OO-KRL
1197	V-55	To OO-RLC/PH-BOZ/OO-LEL
1198	V-56	Preserved at Brussels
1199	V-57	
1200	V-58	
1201	V-59	
1202	V-60	
1203	V-61	
1204	V-62	
1205	V-63	
1206	V-64	
1207	V-65	

Note: A photograph exists of V-4 displaying c/n 1144.

SUD ALOUETTE II

The Alouette helicopter was developed through the early fifties, with the prototype mark II first flying on 12 March 1955. This model is still in production, with over 1,500 sold for civilian and military applications throughout the world. The higher powered Astazou version was introduced in 1961 under the designation SA.318, the earlier aircraft being changed retrospectively from SE.3130 to SA.315.
Belgium received an initial batch of 30 Alouette II, including three for use in the Congo, and these were later augmented by a further twelve. Forty-two Alouette Astazou were purchased for the Landmacht, plus six more for the Rijkswacht, making a total of ninety.
Alouettes are identified by constructors number, and by a production number, which is quoted in the second column below. The serial presentation has varied over the years, beginning in the form 'Axx', then for many

years as 'OL-Axx', and reverting in 1974 to the shorter 'A-xx'. For clarity, the serials shown in column three are in the most modern form, though some of the aircraft concerned will not have survived to carry the serial in this manner. The current 'OTxxx' callsigns are shown in the fourth column, but, for space reasons, only the 'last three' are given.

C/n	Prod.No.	Serial	OT-	Unit/Remarks
1293	127C	A-01	–	w/o 13.9.60
1304	135C	A-02	–	w/o 22.9.60
1305	136C	A-03	AAA	
1341	156C	A-51(No.1)		To Congo AF F-BSGX
1365	166C	A-52(No.1)		To Congo AF
1366	167C	A-53(No.1)		To Congo AF KAT-53
1378	172C	A-04		†
1379	173C	A-05		15Sm/SLV
1422	195C	A-06		18Sm
1423	196C	A-07		†
1467	222C	A-08	AAE	18Sm
1468	223C	A-09	AAF	16Sm
1534	265C	A-10	–	w/o
1535	266C	A-11	AAG	15Sm/SLV
1565	284C	A-12	AAH	15Sm/SLV
1566	286C	A-13	AAI	
1581	294C	A-14	AAJ	15Sm/SLV
1594	301C	A-15	AAK	
1595	302C	A-16	AAL	16Sm
1614	315C	A-17	–	w/o 16.8.67
1624	320C	A-18	AAM	
1626	322C	A-19	–	w/o 5.68
1646	337C	A-20	AAN	18Sm
1661	347C	A-21	AAO	18Sm
1662	348C	A-22	AAP	15Sm/SLV
1666	352C	A-23	AAQ	18Sm
1695	371C	A-24	AAR	15Sm/SLV
1696	372C	A-25	AAS	15Sm/SLV
1710	381C	A-26	AAT	15Sm/SLV
1711	382C	A-27	AAU	15Sm/SLV
1752	409C	A-28	AAV	
1753	410C	A-29	AAW	18Sm
1766	418C	A-30	AAX	
1767	419C	A-31	AAY	ex F-WKQD 15Sm/SLV
1780	428C	A-32	AAZ	15Sm/SLV
1781	429C	A-33	–	w/o
1791	434C	A-34	ABA	16Sm
1792	435C	A-35	ABB	
1802	440C	A-36	ABC	
1803	441C	A-37	ABD	16Sm
1810	444C	A-38		
1851	485C	A-39		

Alouette Astazou

1956	584C-A40	A-40	ACA	16Sm
1958	586C-A41	A-41	ACB	15Sm/SLV
1959	587C-A42	A-42	ACC	
1960	588C-A43	A-43	ACD	15Sm/SLV
1961	589C-A44	A-44	ACE	15Sm/SLV
1962	590C-A45	A-45	ACF	17Sm*
1964	592C-A46	A-46	ACG	16Sm
1986	614C-A63	A-47	ACH	18Sm
1987	615C-A64	A-48	ACI	16Sm
1989	616C-A66	A-49	ACJ	16Sm
1990	617C-A67	A-50	ACK	16Sm
1992	619C-A69	A-51(2)	ACL	16Sm
1993	620C-A70	A-52(2)		†
1995	622C-A72	A-53(2)		†
1996	623C-A73	A-54		†
1998	625C-A75	A-55	ACO	15Sm/SLV
1999	626C-A76	A-56	ACP	16Sm
2009	636C-A85	A-57	ACQ	18Sm
2010	637C-A86	A-58	–	w/o 8.10.69
2018	645C-A94	A-59	ACR	18Sm
2019	646C-A95	A-60	–	w/o
2034	658C-A110	A-61	ACS	18Sm
2035	659C-A111	A-62	ACT	
2049	670C-A125	A-63	ACU	15Sm/SLV
2050	671C-A126	A-64	ACV	17Sm*
2057	676C-A133	A-65	ACW	17Sm*
2064	682C-A140	A-66	ACX	15Sm/SLV
2065	683C-A141	A-67	ACY	
2068	684C-A144	A-68	ACZ	
2072	688C-A148	A-69	ADA	15Sm/SLV
2079	694C-A155	A-70	ADB	17Sm*
2080	696C-A156	A-71	–	w/o 8.8.71
2083	696C-A159	A-72	ADC	15Sm/SLV
2087	700C-A163	A-73	ADD	15Sm/SLV
2094	706C-A170	A-74	ADE	16Sm
2095	707C-A171	A-75	ADF	17Sm*
2110	719C-A186	A-76	ADG	16Sm
2124	730C-A200	A-77	ADH	17Sm*
2133	737C-A209	A-78	ADI	17Sm*
2138	742C-A214	A-79	ADJ	16Sm
2139	743C-A215	A-80	ADK	17Sm*
2142	745C-A218	A-81	ADL	15Sm/SLV
1991	618C-A68	A-90	GCA	Rijkswacht
1994	621C-A71	A-91	–	Rijkswacht w/o
2003	630C-A79	A-92	GCB	Rijkswacht
2004	631C-A80	A-93	GCC	Rijkswacht
2102	712C-A178	A-94	GCD	Rijkswacht
2103	713C-A179	A-95	GCE	Rijkswacht

**1977 'Blue Bees' team*

† Unidentified crashes are either A-04 or A-07 and one from A-52, 53 or 54.

SUD ALOUETTE III

Developed from the Alouette II, the mark III incorporates a larger cabin and covered rear fuselage. Production began in 1959, and the type remains in production, the total built having passed 1,400.

As a partial replacement for the ageing Sikorsky S-58s, the Zeemacht flight at Koksijde - now part of 40 Smaldeel - received three Alouette IIIs in 1971.

C/n	Serial	Callsign
1812	M-1	OTZPA
1816	M-2	OTZPB
1817	M-3	OTZPC

SUPERMARINE SPITFIRE

Without doubt the most famous Allied fighter of the last war, the Spitfire was issued to numerous air forces from surplus RAF stocks from 1945 to 1950. Belgium received a total of 181 aircraft of two marks, plus eleven loaned by the RAF.

Spitfire IX

Also some mark XVI included. The aircraft used by 349 and 350 Smaldelen and the Jachtschool.

RAF loans. Eleven aircraft issued October 1946, but only TB622 and TD188 returned to the RAF. The others involved were TB386, TB709, TB991, TD140, TD184, TD325, TD348, TD372 and RW344.

Serials SM-1 to SM-28. Ex-RAF aircraft, delivered 11 August 1947 to 20 November 1948 –

Ex-RAF	Serial	Remarks
PL349	SM-1	
MJ353	SM-2	w/o 17.5.50
PT887	SM-3	w/o 11.7.50
PL190	SM-4	
TE520	SM-5	w/o 14.2.52
EN568	SM-6	
EN123	SM-7	w/o 18.7.51
PL149	SM-8	w/o 16.10.48
MJ421	SM-9	w/o 4.8.50
TA855	SM-10	w/o 14.12.51
MJ244	SM-11	
MH366	SM-12	w/o 2.9.47
PT643	SM-13	
MJ332	SM-14	
MJ783	SM-15	
PT853	SM-16	R
PT644	SM-17	
ML423	SM-18	
MJ482	SM-19	
PL224	SM-20	w/o 1.9.52
MK577	SM-21	R
MJ559	SM-22	
MJ617	SM-23	w/o 25.6.52
MK777	SM-24	R
TA836	SM-25	
PV189	SM-26	w/o 25.6.52
RK851	SM-27	w/o 22.11.51 R
MK153	SM-28	R

Aircraft marked 'R' re-serialled SM-44 to 48 see below.

Serials SM-29 to SM-43. Aircraft originally supplied ex-RAF to the Netherlands East Indies Air Force, and delivered after overhaul by Fokker, June 1952 to 1953. The 'B-' serials in the fourth column below were test-flight serials used by Fokker. Withdrawn 1956.

Ex-RAF	Serial	Ex-Dutch	Test	Remarks
MK912	SM-29	H-119,H-59	B-1	
MH485	SM-30	H-102,H-51	B-2	w/o 29.3.53
MK205	SM-31	H-102,H-52	B-3	w/o 11.10.52
MH424	SM-32	H-106,H-53	B-4	
MH439	SM-33	H-120,H-56	B-5	
NH309	SM-34	H-113,H-57	B-6	

Hawker Hurricane IIc LF345 was delivered on 2 June 1946 and coded ZA-P (call-sign OTZAP). Following a sojourn in the Musée de l'Armée as ML-B in an incorrect camouflage scheme, it has recently been restored to its correct colours. *(Tony Lloyd)*

Gloster Meteor F.8 EG224
K5-K was the first of 37 such aircraft built by Avions Fairey from Gloster-supplied components. It is shown in preserved condition at Brustem in 1968. *(P. Glas)*

Gloster Meteor F.8 EG234
B2-0 awaits disposal at Koksijde during the mid-sixties. It last served with the target facilities flight at Koksijde. *(D.M. Sargent)*

Gloster Meteor NF.11 EN-23
Twenty-four Meteor NF.11 all-weather fighters replaced a similar number of Mosquitoes with 1 Wing in 1952-56, but were withdrawn in favour of the Canuck in 1958. EN-23 became OO-GEZ for a short time after disposal in August 1958. *(MAP)*

Gloster Meteor NF.11s at Ostend Airport in August 1960 showing the civilian identities crudely applied on disposal. They are EN-2/KT-E/OO-ARR, EN-21/ND-J/OO-ARP, EN-5/KT-S and EN-11/KT-T. *(B.J. White)*

Hawker Hunter F.6 IF-137 of the Diables Rouges/Rode Duivels team, before replacement by the now familiar Magisters. The colour scheme was red, similar to the current one. *(MAP)*

Hawker Hunter F.6 ID-126
7J-H of 7 Smaldeel, 7 Wing, was retired in 1962, and ferried to Hawker Siddeley for refurbishing, and later re-sold to Iraq. *(MAP)*

Hawker Hunter F.4 ID-44 photographed at Brustem in 1968 served with the Rode Duivels/Diables Rouges, before storage at Zellick, and is now preserved in Brussels. *(W. J. Bushell)*

Hawker Hunter F.6 IF-48 was built by Fokker and assembled in Belgium by Avions Fairey. After service with 26 Smaldeel/9 Wing it went via Hawker Siddeley to Iraq. *(MAP)*

Hawker Hunter F.6 IF-61 is identified as belonging to 8 Smaldeel by its 'OV' code prefix. These markings were deleted later in the Hunter's career, and replaced by the numerical part of the serial. *(MAP)*

Lockheed T-33A FT-6 served with the Jachtschool in the late fifties, and is seen here with the unit badge on the fin. Between 1962 and 1972 it served with the Dutch Air Force as M-44, but has now returned to its former serial with 11 Smaldeel to await replacement by the Alpha Jet. *(MAP)*

Lockheed T-33A FT-22 of 11 Smaldeel, seen at Kleine Brogel in June 1976, shows that camouflage has spread to the T-33A as well as the Belgian Air Force's tactical aircraft. *(S.G. Richards)*

Hawker Siddeley HS.748-2A CS-03 shown at Farnborough in September 1976 prior to delivery to 21 Smaldeel; it was the last of three to be delivered for short range transport. *(S.G. Richards)*

Lockheed C-130H Hercules CH-10 of 20 Smaldeel was delivered to Melsbroek in natural metal finish, but like its companions was flown almost immediately to Marshalls at Cambridge for camouflaging in March 1973. *(MAP/Ian MacFarlane)*

Lockheed F-104G Starfighters FX-90 and FX-21 take off from Alconbury in June 1975. Although the Starfighter is hardly the the ideal aerobatic display type, they were used by 10 Wing's 'Slivers' team. *(S.G. Richards)*

Lockheed F-104G Starfighter FX-04 was the second Belgian-built example to be delivered to the BLu, on 18 April 1963. Its camouflage colours and small white serial are representative of the present markings style for the type. *(W.J. Bushell)*

Lockheed F-104G Starfighter FX-19 at Woodbridge, August 1976. 1 Wing aircraft (349 and 350 Smaldelen) operate in the air defence role and are equipped to carry Sidewinder missiles on the wingtip. *(Cambridge Aviation Photographers)*

Lockheed F-104G Starfighter FX-78 of 31 Smaldeel, 10 Wing demonstrates the practice prior to the adoption of central servicing of carrying the squadron badges by the cockpit side, and wing insignia on the fin. It also has the original style large black serial. *(V. Kenens)*

Lockheed F-104G Starfighter FX-12 seen at Bierset in June 1969, was originally delivered with the serial FX-012, and is one of a number of aircraft which, shortly after camouflaging, flew with large white serials. *(W. J. Bushell)*

MJ714	SM-35	H-117,H-67	B-7	w/o 19.3.54
NH238	SM-36	H-103,H-60	B-8	COGEA
MK923	SM-37	H-104,H-61	B-9	COGEA
MH725	SM-38	H-112,H-63	B-10	
NH188	SM-39	H-109,H-64	B-11	COGEA
MH415	SM-40	H-108,H-65	B-12	COGEA
MH434	SM-41	H-105,H-68	B-13	COGEA
MJ893	SM-42	H-110,H-69	B-14	
MK297	SM-43	H-116,H-55	B-15	COGEA

The aircraft marked 'COGEA' above were transferred to that firm for target-towing duties, and later sold abroad as follows–

SM-36	OO-ARE N238V,'EN398'
SM-37	OO-ARF N93081, NX521R
SM-39	OO-ARC CF-NUS
SM-40	OO-ARD G-AVDJ, N415MH
SM-41	OO-ARA G-ASJV
SM-43	OO-ARB G-ASSD, N1882

Serials SM-44 to SM-48 were allocated to five earlier aircraft, presumably after conversion. They were previously SM-16, SM-21, SM-24, SM-27 and SM-28, but in which order is not known.

Spitfire XIV

One hundred and thirty-three aircraft (including instructional airframe RM862/6350M) from RAF stocks, deliveries commencing 17 April 1947. The initial order was for 100, but SG-39 and one other were refused by the Belgian authorities as unsuitable, and replaced by SG-101 and 102. Spitfire XIVs were issued to 349, 350 1, 2 and 3 Smaldelen, and to the Jachtschool.

Serials SG-1 to SG-25 delivered 1947 –

Ex-RAF Serial	Serial	Codes	Remarks
NH655	SG-1	GE-N	
RB166	SG-2	GE-B	
RM916	SG-3	GE-A,UR-M	
MV248	SG-4		
RB163	SG-5	GE-O	
RM701	SG-6	GE-E	
RN121	SG-7	3R-G	w/o 15.3.51
NH654	SG-8	UR-O	
SM938	SG-9	GE-O, IQ-S	w/o 12.5.50
RM768	SG-10	IQ-G	w/o 14.4.48
RM672	SG-11	3R-T	w/o 5.8.53
RM870	SG-12	MN-H,IQ-Z	w/o 5.1.50
RB154	SG-13	MN-K,GE-E IQ-V	
RM726	SG-14	UR-R,GV-R	w/o 8.9.52
RM685	SG-15		
NH688	SG-16	IQ-L	w/o 3.7.50
RM935	SG-17		
RM741	SG-18	MN-A	
RN206	SG-19	YL-A	
NH712	SG-20		w/o 16.1.48
RM676	SG-21		w/o 19.2.48
RM791	SG-22	IQ-O	
RM679	SG-23	IQ-C	
RM876	SG-24	MR-B,IQ-B	w/o 27.12.50
RM927	SG-25	IQ-W,3R-A	

Serials SG-26 to SG-80 all delivered 1948 –

NH643	SG-26		w/o 23.7.51
RB161	SG-27	MN-P,IQ-O	w/o 31.7.52
RM787	SG-28		w/o17.3.48
RN113	SG-29		
MV288	SG-30	IQ-K	w/o 10.8.51
RN201	SG-31	MN-L,YL-B Preserved 'GE-A'	w/o 5.2.50
RM759	SG-32	GE-M,YL-C	
TZ137	SG-33	MN-B	w/o 5.6.51
MV381	SG-34	GE-B	
RM770	SG-35	MN-G,YL-F	w/o 5.6.51
MV302	SG-36		
RM860	SG-37		w/o 14.1.49
RM764	SG-38	GE-A	w/o 24.2.52
MV256	SG-39		
RB186	SG-40	YL-S	
RM866	SG-41		w/o 30.4.48
RM697	SG-42	GE-D,IQ-Z	w/o 4.7.50
RM920	SG-43	MN-C,3R-J, YL-A	w/o 29.2.51
RM802	SG-44	IQ-U	w/o 23.12.53
RN119	SG-45	IQ-Q,IQ-W	
RM625	SG-46	UR-G	w/o 14.1.49
RN117	SG-47	IQ-J	w/o 9.8.51
TZ132	SG-48		
TX995	SG-49		
MV378	SG-50	3R-A,8S-R IQ-BG	
SM821	SG-51	UR-F	w/o 5.5.49
RM882	SG-52	IQ-S,3R-V	w/o 16.3.51
RM857	SG-53	3R-O	w/o 13.6.49
RN215	SG-54	IQ-K	
MV246	SG-55		w/o 11.10.48
NH754	SG-56	IQ-D,MN-W	w/o 5.6.50
RM921	SG-57	RL-D	
RM680	SG-58	IQ-DD	
RM710	SG-59	MN-V,3R-N	w/o 6.4.51
RM933	SG-60	GE-S,3R-V	
RM822	SG-61		
RM913	SG-62	GE-P	
RN115	SG-63	GV-R	w/o 31.7.51
TZ192	SG-64	GE-P,IQ-F	w/o 23.10.52
NH892	SG-65	MN-X,3R-C, GV-X,IQ-M	w/o 14.12.53
RM705	SG-66	MN-V,IQ-O	w/o 3.53
NH864	SG-67	MN-Y,YL-D	w/o 28.3.51
RB156	SG-68	GE-F	w/o 14.5.48
RB182	SG-69	GV-F,IQ-U	
RM683	SG-70	UR-C	
NH718	SG-71	IQ-A	
RM841	SG-72	UR-A	w/o 30.6.49
NH720	SG-73	3R-A,UR-A	w/o 14.6.51
NH780	SG-74	GE-R	
NH797	SG-75	MN-N	
RM917	SG-76		w/o 25.2.53
MV263	SG-77	IQ-T	w/o 18.7.50
RM703	SG-78	UR-E,GV-E	w/o 20.10.50
TZ166	SG-79		w/o 20.9.50
MV382	SG-80	MN-T,GV-J, GV-S	w/o 20.7.52

Serials SG-81 to SG-102, delivered up to 21 July 1949 –

TZ127	SG-81	
RM937	SG-82	
RM820	SG-83	MN-E
NH857	SG-84	3R-O
TZ193	SG-85	
RM863	SG-86	GE-K
NH807	SG-87	
TX989	SG-88	GE-C
TZ111	SG-89	MN-M
MV369	SG-90	IQ-B
NH775	SG-91	YL-G
RM938	SG-92	
NH863	SG-93	MN-G,IQ-D
NH894	SG-94	MN-L
MV383	SG-95	3R-D
RN124	SG-96	
TZ154	SG-97	
NH831	SG-98	
MV312	SG-99	
RM784	SG-100	
NH922	SG-101	
RN116	SG-102	IQ-C

Serials SG-103 to SG-132 delivered via Vickers in 1951 under Western Union contract. With the odd exception*, the ex-RAF and BAF serial tie-ups are not known but the aircraft concerned were –

MV265	MV267	MV359	NH658	NH702
NH710	NH741	NH742	NH743	NH789*
NH798	NH838	NH904	NH918	RB165
RM674	RM700	RM707	RM790	RM792
RM795	RM879	RM918	SM930	TX992
TZ142	TZ174	plus three unknown		

A few of the serial/code tie-ups are known:

Ex-RAF	Serial	Code
	SG-104	IQ-N
	SG-105	IQ-R
NH789	SG-108	IQ-V, B2-K
	SG-112	IQ-Y
	SG-114	GV-R
	SG-117	IQ-M
	SG-120	YL-P, IQ-D
	SG-121	IQ-S
	SG-125	IQ-W

SWEARINGEN MERLIN 3A

Prior to joining forces with Fairchild, the Swearingen company specialised in the conversion of other manufacturer's types to executive standard. The prototype Merlin, which used a new fuselage mated to a Queen Air wing, was flown 13 April 1965, and the 8-seat Merlin 2A entered production. The elongated Series 3 8-10 seater appeared in 1970.

The Belgian Air Force ordered six aircraft in 1975 as replacements for the Percival Pembrokes with 21 Smaldeel. Deliveries were made between 28 March 1976 and September 1976.

Serial	Callsign
CF-01	BAF51
CF-02	BAF52
CF-03	BAF53
CF-04	BAF54
CF-05	BAF55
CF-06	BAF56

WESTLAND SEA KING MK.48

A licence built version of the Sikorsky S-61, the Sea King began production in 1969, primarily as an anti-submarine helicopter for the Royal Navy, but also in an SAR version for the RAF, and as the Commando assault helicopter supplied to Egypt and Qatar. Other exports have been to Australia, West Germany, India, Norway and Pakistan.
Five Belgian aircraft ordered 22 April 1974 to replace the Sikorsky S-58 with 40 Smaldeel. Crew training was carried out at Culdrose by the Royal Navy Foreign Training Unit, with deliveries made to Culdrose between 20 May (RS-02) and 14 July 1976 (RS-04). The training unit disbanded on 5 November 1976, following which the aircraft transferred to Koksijde for 40 Smaldeel.

C/n	Serial	Ex-
WA831	RS-01	G-17-1, G-BDNH
WA832	RS-02	G-17-2, G-BDNI
WA833	RS-03	G-17-3, G-BDNJ
WA834	RS-04	G-17-4, G-BDNK
WA835	RS-05	G-17-5, G-BDNL

APPENDIX A

Belgian Military Aircraft Orders [pre-1940]

In approximate order of delivery –

Farman HF3
Farman/Bollekens HF3
Farman/Bollekens HF20
Bollekens Jero
Deperdussin Monocoupe
Bleriot Artillerie
Morane 1911
Bleriot Genie
Farman/Bollekens F16
Bleriot Olieslagers
Farman MF7/F56/F60
Farman HF21
Farman F40
Farman F41
Farman MF13
Farman MF14
Nieuport 10
Ponnier
Voisin Canon
HD 1
Short 225D
Nieuport 12C/16C
Nieuport 17C
Schreck Type 2
Caudron G3
Sopwith Pup/Camel

Spad 11/13/16
BE 2C
RE 8
Breguet XIV
Letord 1
Friedrichshafen G3
Salmon 2A2
Fokker DVII
Gotha G3
Rumpler CVI
LVGC-6
Siemens SSWD-III
Aviatic
Halberstadt CL
Avro 504K
Avro/SABCA 504K
Avro 504N
Avro/SABCA 504N
Martinsyde F6
Central Centaur
Bristol Brisfit
Bristol 300
MS35AR
DH.4
DH.9
DH.60 Moth

Fiat CR10
Nieuport Delage 29C-1
Nieuport/SABCA 29C-1
Avia/SABCA BH21
Avia BH33
Nieuport Delage 72
MS 230
Breguet/SABCA Br19
Ansaldo A-1
Renard 32
Renard 31
Renard 26
Renard 22
Renard 23
Renard Epervier
Ansaldo A300
Ansaldo/SABCA A300
Cierva C30A
Fairey Firefly
Fairey Fox 2/3
Fairey Fox VI
Fairey Fantome
Koolhoven FK56/57/58
Gloster Gladiator
Fiat/SABCA CR42
Fairey Battle

Hawker Hurricane
Caudron Goeland
Nieuport 81
Potez 33/36
Fokker F7
Breguet Br694
LACAB Gr 8
Leo 45
Renard R38
SV-4
SV-5
SV-10
SM83
Bulté RB

Orders outstanding 1940

Brewster Buffalo
Caproni/SABCA SA47
Breguet 694

APPENDIX B

Belgian Squadrons of the R.A.F.

During World War 2, large numbers of Europeans who managed to escape to Britain ahead of the advancing German armies formed their own units within the three services. Two Netherlands Navy squadrons were integrated with RAF Coastal Command but there were insufficient Belgian and Dutch Air Force crews to immediately form new RAF squadrons and so these escapees were distributed amongst the RAF Commands or made up one Flight within certain units. Eventually, however, the two countries provided the bulk of three Spitfire squadrons, one Mitchell bomber squadron, and one Catalina patrol squadron, in the RAF and one convoy defence squadron of Swordfish in the FAA.

Arriving, as they did, at the lowest ebb of Britain's fortunes the Belgian and Dutch Contingents performed an inestimable service for all the people of free Europe in lending their talents to the RAF when they were most desperately needed. Particular mention must be made of those Belgians who disregarded orders to surrender issued by their own government and chose to find their way to Britain at a time when the world was giving her little more time than their home country to resist the Nazis once they had landed. There is little wonder that the RAF squadron numbers are perpetuated today as part of the two national air forces as an inspiring reminder of former battle honours.

Histories of the squadrons are given below:

349 (Belgian) Squadron, RAF

The squadron was raised at Ikeja, West Africa in January 1943 (its first officer arriving on the 9th of that month) and began to equip with the Curtis Tomahawk with the intention of serving in the Belgian Congo. These plans were soon abandoned, and on 17 March, 349 squadron became a ferry unit for the transfer of aircraft from Takoradi to the Middle East, disbanding in late April.

Under the command of Flt/Lt. I. G. du Monceau de Bergendal, DFC, 349 Squadron re-formed at Wittering on 5 June 1943, receiving its first pilot three days later. Seven more Sergeant Pilots arrived from 350 and 610 Squadrons on the 13th, and four days later the squadron flew its first aircraft in Britain – a Ju 88 on loan from the Enemy Aircraft Flight, based at Wittering. The arrival of the first pilot had coincided with the transfer of 349 Squadron to Collyweston (a satellite of Wittering), and when the first Spitfire arrived, it touched down here on the 22nd. The other Wittering sub-station at Kingscliffe received the squadron on June 29th on which day the second Spitfire was delivered. With a full complement now on charge, the unit transferred to Wellingore on 5 August, becoming operational there on the 13th and to Acklington on the 25th (the day after its first operation). The training period over, 349 received the Spitfire Vb and Vc prior to moving to Friston on October 21st. As part of the invasion build-up, the unit, newly equipped with the Mk.IX, moved to Hornchurch on 11 March 1944; 135 Wing, Selsey (with 222 and 485 Squadrons) on 11 April and, as units were rapidly transferred during the immediate post-invasion period, to Coolham (30 June), Funtington (4 July) where S/Ldr. Van der Velde assumed command and Tangmere (19 August). At last on 31 August, 349 Squadron reached the Continent, landing at airfield B17 (Carpiquet). A rapid allied advance resulted in transfers to B35/Godelemensil (8 September), B53/Merville (12 September), B65/Maldegem (2 November), B77/Gilze-Rijen (13 January 1945).

Now poised for the final stroke against Germany, 349 was instead withdrawn from Gilze-Rijen on 19 February and transported to the peaceful landscape of Cornwall where it arrived at Predannack three days later. Any initial disappointment was dispelled by the delivery of three Tempests on 24 February and a similar number a few days later. Two Typhoons were additionally taken on to act as transitionary trainers between the Spitfires and Tempests, but before any more of the latter arrived, the re-equipment programme was cancelled on 3 March.

Between 16 and 19 April the unit (now under S/Ldr Lallemand) made its way to B106/Twenthe and on 21st took over 331 (Norwegian) Squadron's complement of Spitfires. A posting to B113/Varrelbusch followed on 30 April and it was here that 349 Squadron made its last operational flight on 4 May and a few days later celebrated the coming of peace.

On 29 June, the unit transferred to B116/Wunsdorf, being joined shortly afterwards by sister squadron 350. Both made a brief visit home to Melsbroek between 19 and 22 July 1945, and a similar excursion on 1 to 4 September. As part of the forces of occupation, 349 transferred to B152/Fassberg on November 29. During December S/Ldr. van der Velde again took command and in preparation for transfer back to Belgium, an advance party left for Brustem, on 30 April 1946 whilst national roundels were applied to the aircraft at Fassberg in addition to the tricolour insignia which had been carried on the side of the cockpit of each aircraft during the war.

The move to Brustem took place on 2 May, but a month later the unit was back at Fassberg taking a week off (8 to 15 June) for an armament practise visit to Sylt. During July squadron strength and servicability dropped dramatically with the demobilisation of 75% of the personnel. None-the-less 349 pulled through and was itself 'demobbed' on 24 October when it officially left the RAF to become a Belgian unit in full.

Examples of 349 Squadron aircraft:

Tomahawk I

AH775/GE-N	AH805 (12-3-43)	AH923
AH972		

Kittyhawk II

FS401 (9-4-43)	FS403	FS415 (31-3-43)
FS416		

Harvard III

EX500 (at Ikeja)

Spitfire V

W3262/GE-C	W3373/GE-X	AA751/GE-K
AA765/GE-I	AB175/GE-U (11-2-44)	AB202/GE-B
AD192/GE-S	AD295/GE-P	AR292/GE-T
AR293/GE-Y	AR432/GE-G (12-2-44)	AR437/GE-J
AR490/GE-A (11-2-44)	AR492/GE-T	BL334/GE-F
BL385/GE-O	BL565/GE-E (3-1-44)	BL642/GE-C1
EE660/GE-Z	EE745/GE-V to GE-Z	EN781/GE-H (3-5-44)
EP354 (12-2-44)	EP334/GE-L	EP660/GE-Q

Spitfire IX

MH371/GE-L	MH434	MH610/GE-Z
MJ230/GE-K	MJ253/GE-M	MH294/GE-X
MJ353/GE-J	MJ363/GE-U	MJ369/GE-H
MJ502	MJ748/GE-A (7.6.44)	
MJ787	MJ879/GE-G	MJ889/GE-W
MJ955/GE-T	MJ962/GE-F	MJ964
MK130/GE-P (28-5-44)	MK135/GE-P	MK136/GE-N (15-4-44)
MK148/GE-R	MK153/GE-S	MK175/GE-B (12-6-44)
MK178	MK192/GE-H (21-5-44)	MK230
MK233/GE-C	MK249	MK252 (8-6-44)
MK302	MK354/GE-V	MK362/GE-E,
MK363	MK368/GE-U	MK663
MK717	MK787	MK804
MK830	ML365	ML404
ML407	NH199	NH241
NH257	NH261	NH311
NH323	NH354	NH381
NH419	NH423	NH425
NH426	NH434	NH453
NH454	NH456	NH464
NH472	NH474	NH484
NH485	NH538	NH589
PL128	PL139	PL152
PL161	PL451	PL492
PT360	PT391	PT395 (6-10-44)
PT549 (3-2-45)	PT555 (19-10-44)	PT559
PT718	PT723	PT730 (5-10-44)
PT763	PT830	PT841 (19-10-44)
PT851	PT891	PT946
PV134 (3-11-44)	PV155	PV175
PV310	RK802	RK803
RK809	RK817	RK851
RK853	RR250	SM181
SM186	TA837 (5-2-45)	TB581
TB613	TB735	TB906
TB910	TB991/GE-L	TD121 (13-2-46)
TD140	TD184	TD288
TD372/GE-B	TE191/GE-G	TE274/GE-K
TE349/GE-T	TE444/GE-P	

350 (Belgian) Squadron, RAF

First of the Belgian units to gain autonomy within the RAF, 350 Squadron formed at Valley on 12 November 1941 out of the Belgian Flight of 131 Squadron, receiving its first three Spitfires P7613, P8020 and P8652 on the 25th of that month, having S/Ldr. J. M. Thompson as its first C.O. With nineteen aircraft (excluding a communications Magister) on charge the unit became operational, mostly for convoy protection duties, on 22 December, adding a further seven machines before moving to Atcham on 19 February 1942, On 5 April, 350 Squadron proceeded to Warmwell for armament practise, having replaced its initial Spitfire IIa with more recent Vb examples, moving straight to Debden when the course was completed on 15 April and shooting down its first German aircraft (an FW 190) on 9 May. From then until the end of the war, the squadron was constantly moving bases: Gravesend 30 June, Martlesham 7 July for gunnery training, Kenley 16 July, Redhill 31 July, Southend 23 September, Hornchurch 7 December, Heston 1 March 1943, Debden 5 March. A period of 'rest' in Northern England during the spring and summer included residence at Acklington 1 April, Ouston 8 June, back to Acklington 20 July, Digby 25 August before returning South to West Malling on 7 September. A short detachment to Digby for two days (18 and 19 September) preceeded a move to Hawkinge on 1 October and Hornchurch on 30 December where 350 Squadron passed on its Spitfires to take over those of 222 Squadron. Returning to Hawkinge on 10 March 1944, the unit made immediate preparations for a transfer to Peterhead which it executed on 13 and 14 March. On 25 April it returned to the operational area at Friston with Mk.Vc Spitfires from where, together with 501 Squadron, it was attached to Air Defence of Great Britain during the Invasion.

The squadron received the Spitfire IX in July and the XIV in August but remained in Britain, albeit on the front line until December flying from West Hampnett 4 July, Hawkinge 8 August and Lympne 1 October. At last, on 3 December it made the long awaited move across the Channel to B56/Evere in its home country and to the US airfield Y32/Zwartberg on 1 January 1945. Eindhoven (27 January) was the next stop and with the exception of a gunnery practise exercise at Warmwell between 18 March and 2 April, it remained there until transferred to B106/Twenthe on 7 April arriving in Germany at B118/Celle on the 16th. As the final shots of World War 2 were being fired, the unit took up residence at B152/Fassberg on 6 May as part of 122 Wing of the Occupation Forces.

During July the squadron made a short visit to B172/Husum before settling down at B116/Wunsdorf from where it made a triumphant four day return to Brussels/Melsbroek on 19 July. Thereafter the squadron returned to Wunsdorf, but moved to Fassberg the next month, remaining there as part of the occupation forces until taken off the RAF order of battle on 24 October 1946, by which time it was back in Belgium under the command of S/Ldr. van Lierde.

Examples of 350 Squadron aircraft:

Magister	**Tiger Moth**
T9804	DE484

Spitfire II		
P7297/MN-A	P7376	P7378
P7509	P7620	P7826
P7900	P7976	P8020
P8033	P8040	P8089
P8091	P8144	P8146
P8149	P8200/MN-T	P8241
P8377	P8652	

Spitfire V		
P8702 (9-1-42)	R6809	W3209
W3214	W3261	W3518
W3626/MN-S (1-6-42)	W3646	W3759
W3843	W3898/MN-J	W3899
W3970/MN-D	AA729/MN-B	AA833/MN-E
AA853/MN-C	AA910	AA915
AA933	AA934	AA945/MN-K
AA978	AB173 (1-6-42)	AB180
AB183	AB276	AB380
AB492	AB818	AB875 (21-3-44)
AB912	AB931 (21-3-44)	AB967
AB971	AB989	AD113
AD228	AD248	AD288
AD314 (20-12-43)	AD322 (5-8-42)	AD428
AD469	AD472	AD475
AD550 (12-12-42)	AD573	AR366
AR373	AR380 (19-8-42)	AR393
AR448	AR451	AR452
AR492/MN-M	AR498/MN-G (30-4-44)	AR515
AR592	BL291	BL370
BL464	BL476 (23-5-42)	BL496/MN-O
BL540	BL563	BL621
BL622 (19-5-42)	BL631	BL696
BL780	BL822 (1.6.42)	BL907
BL936 (1.6.42)	BL979	BM176/MN-F
BM240	BM272	BM343/MN-W
BM344/MN-R	BM363 (8-6-44)	BM381
BM422 (10-6-44)	BM468 (28-12-43)	BM474
BM513	BM564	BM583
BM652 (13-11-43)	EE613/MN-E	EE723/MN-E (14.6.44)
EE726	EE727	EE738
EE739 (28-3-43)	EE743	EE744
EE753	EE766/MN-C	EN769 (13-8-42)
EN781	EN794	EN796
EN800	EN854/MN-H (29-3-44)	EN950 (6-6-44)
EN957	EP240 (13-11-43)	EP277
EP288	EP387	EP491
EP640 (28-3-43)	EP664	EP687
EP699		

Spitfire IX

MH371/MN-J	MH431/MN-P	MH423/MN-K
MH428 (30-1-44)	MH430/MN-E	MH432/MN-A
MH434/MN-B	MH439/MN-V	MH476 (30-1-44)
MH491/MN-U (10.5.44)	MH496/MN-Q	MH499/MN-T
MH753/MN-O	MJ118/MN-H	MJ150/MN-G
MJ201/MN-	MJ232/MN-C	MJ243
MJ253/MN-M	MJ296/MN-G (29-4-44)	MJ338
MJ416/MN-D	MJ423	MJ521
MJ790	MJ890	MJ900
MK192/MN-Z (21-5-44)	MK193	MK208 (16-9-44)
MK265	MK298/MN-B	MK319
MK320	MK343	MK520
ML137	ML143	ML294

Spitfire XIV

NH654/MN-X	NH658/MN-D	NH659/MN-F (27-7-45)
NH660/MN-S	NH661/MN-Y	NH686/MN-V (20-4-45)
NH689/MN-B	NH690	NH692
NH693/MN-J	NH695	NH696
NH697/MN-K	NH698/MN-C	NH702
NH705	NH710	NH711 (23-1-45)
NH714	NH716 (9-11-44)	NH718
NH720	RB152	RB154/MN-N
RB155/MN-O	RB158/MN-L	RB159
RB161	RB168	RB169/MN-F
RB173	RB176	RB177
RB180/MN-Z	RB181/MN-H	RB183/MN-Z
RB185	RB186/MN-W	RB188
RB189/MN-G	RM618/MN-P	RM619 (16-1-45)
RM620	RM621	RM622 (1-1-45)
RM623/MN-A	RM648/MN-R	RM655
RM671 (14-11-44)	RM672	RM673 (25-12-44)
RM675	RM690 (24-12-44)	RM691
RM695	RM696	RM700
RM701	RM728	RM729/MN-M
RM733/MN-H	RM741	RM744
RM747	RM748	RM749
RM750	RM752	RM753
RM754	RM755	RM756
RM760	RM761	RM762
RM763	RM764	SM825/MN-M

Spitfire XVI

TB866/MN Q	TB877/MN-G	TD231/MN-J
TD281/MN-S	TD325/MN-M	TE119/MN-H

Dates in parentheses are those on which the aircraft was written off whilst in service with that squadron.

RAF (Belgian) Training School

On 8 November 1943, the RAF Belgian Depot at Goring-on-Thames was re-named the RAF (Belgian) Training School under the command of Squadron Leader R. Cajot. Due to a shortage of accommodation, however, some of the personnel were billited at Kidlington, until on 1 January 1944 the unit was split, with the Depot section remaining at Goring and the Reception & Training section at Snitterfield under Wing Commander D. A. Guillaume within 54 Group. The latter was responsible for the initial training of air- and ground crew, but at this stage, Officers and Men were passed to the Flying Training and Technical Training schools of the RAF to receive their specialist instruction.

The school began a move to Snailwell on 10 October 1944, and completed the transfer three days later when it came under 28 Group, Technical Training Command. At last, on 10 February 1945, the unit began to receive its first aircraft: five Miles Masters arrived from 5(P)AFU on that day, followed by nine on the 13th and five the next day. No time was lost in commencing instruction and the first lessons began on 17 February. Twelve Tiger Moths arrived from Maintenance Units on the 20th, and on the 21st, a further Master and two Dominies (NR776 and NR777) were added.

Training continued during March, but it was not until the 10th April that the school was officially opened by Group Captain Guillaume, who was by this time Deputy Director General of the fledgling Belgian Air Force. The first solo flights were made by pilots Tacquin and Born on the 17th, and during the month the school accumulated 120.45 hours dual and 28.55 hours solo. May 1945 saw these totals increased to 358.50 and 322.45 respectively, with the result that recruit reception was transferred to Bottisham in June to make more room at Snailwell, and by August personnel strength had risen to 179 officers, 224 SNCOs and 1400 Corporals and airmen.

During 1946, the various elements of the Training School were transferred to Belgium as noted in the main unit histories section under Centrum voor Militaire Vorming, Elementaire Vliegschool and Technische School van de Luchtmacht.

APPENDIX C

Belgian Military Airbases

BELGIUM

1 Bevekom
2 Bierset
3 Brasschaat
4 Brustem
5 Chièvres
6 Florennes
7 Goetsenhoven
8 Kleine Brogel
9 Koksijde
10 Melsbroek

BELGIAN CONGO

1 Kamina
2 Leopoldville (now Kinshasa)

LUXEMBOURG

11 Luxembourg

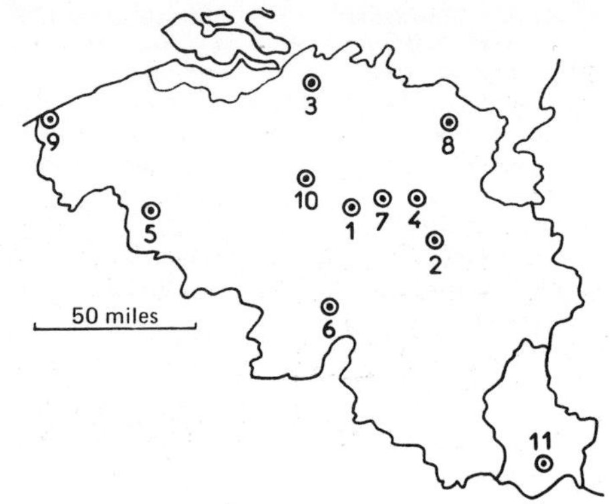

Other airfields used by military aircraft include -

Gosselies	North of Charleroi : Avions Fairey / SABCA overhaul of Starfighter and Mirage aircraft.
Weelde	North of Turnhout : Air Cadet gliding school with Super Cub tugs.

Note:

7104 Air Base Squadron / 86TFW, USAF is resident at Chièvres, north of Mons. The based VC-118A (0-33303) is for the use of VIPs visiting the NATO HQ near Brussels.

Lockheed TF-104G Starfighter FC-06
Most of the ten remaining Lockheed-built TF-104G two-seaters now serve with 10 Wing, whose FC-06 with lion insignia on the fin is seen in August 1970. *(S.G. Richards)*

Lockheed F-104G Starfighter FX-46
formerly with 1 Wing/Slivers, was written off in January 1976. *(E. Moreau)*

North American Harvard 2A H-39 is now in the Musée de l'Armée. *(P. Glas)*

Percival Proctor IV P-4 ex-RAF NP171 was withdrawn following an accident in May 1949, and thus available for preservation. It is displayed in the Musée de l'Armée in the colours which it wore whilst serving with 367 Smaldeel. *(Tony Lloyd)*

Percival Pembroke C.51 RM-2 OT-ZAB Note the early colours. *(Percival Aircraft)*

Percival Pembroke C.51 RM-5 OT-ZAE, at Birmingham shortly before being retired in favour of the Merlin in 1974. *(S.G. Richards)*

Piper L-18C Super Cub OL-L25 of 15 Smaldeel, Belgische Landmacht, although like most of the Army aircraft, bearing no unit markings. After retirement from military flying, it went on to enjoy a further lease of life as a civilian aircraft in Germany. *(P. Glas)*

Piper L-18C Super Cub L-87 photographed at Kassel in April 1960 with its serial in the same position as the Austers, pre-dating the deletion of the 'OL-' prefix by fourteen years. It later became OL-L87, and is now preserved in Brussels. *(MAP)*

Republic F-84G Thunderjet FZ-132 YL-D has the holly-leaf badge of 3 Smaldeel on the nose, and the divided diamond of its parent unit, 2 Wing, on the fin. The motto above the leaf is *'Qui s'y frotte s'y pique'*, a declaration shared with SPA38 of the French Air Force. *(W. J. Bushell)*

Republic F-84E Thunderjet FS-2 has been preserved as a gate-guardian at Kleine Brogel, hence the fictitious code. Both Holland and Belgium received 21 before the improved F-84G was delivered in quantity. *(S.G. Richards)*

Republic F-84F Thunderstreak FU-32 of 3 Smaldeel, 2 Wing, displays the squadron code letters applied to the type during the earlier part of its Belgian service. *(MAP)*

Republic F-84F Thunderstreak FU-58 has white wings and fin, for use in an aerobatic display. When codes were discontinued, Thunderstreaks were marked on the nose with a three-figure number corresponding to the serial. *(via S.G. Richards)*

Republic RF-84F Thunderflash FR-9
The first Thunderflashes delivered to 42 Smaldeel were coded from H8-A onwards in serial order, with the exception of H8-I. FR-9 was withdrawn to Koksijde in 1971 when the squadron re-equipped with the Mirage 5BR. *(MAP)*

Republic F-84F Thunderstreak FU-193 coded 3R-N of 1 Smaldeel, 2 Wing, photographed at Chièvres in April 1960, one month before it was written off in a crash. *(MAP)*

Republic F-84F Thunderstreak FU-106 at Alconbury in July 1969 wearing grey/green/azure blue camouflage. *(S.G. Richards)*

Republic RF-84F Thunderflash FR-34 was the last of four Thunderflashes transferred from the Luftwaffe to 42 Smaldeel in 1965, and is seen here at the Alconbury Armed Forces Day display on 14 August 1971, bearing the inscription *'Adieu a toi, le petit dernier'* - *'Goodbye to you, little last one'*, a reference to it being the last of the type overhauled by Avions Fairey. *(D.M. Sargent)*

RF-84F Thunderflash FR-22 and FR-18 In the foreground, at Deelen in May 1970, FR-22 is seen wearing the insignia of its KLu counterpart, 306 Squadron, although the latter had by that time foresaken its RF-84F's for Starfighters. *(via P.A. Jackson)*

SIAI SF.260M ST-25 at Kleine Brogel in June 1976. All thirty-six SF.260Ms were received by the EPE/EVS at Goetsenhoven, and the school's penguin badge is displayed on the fuselage. *(S.G. Richards)*

Stampe SV-4c V-46 at Coltishall in September 1966 also displays the EPE/EVS penguin badge, and illustrates their earlier equipment. *(Cambridge Aviation Photographers)*

Sikorsky S-58 B-12 OT-ZKL was one of seven ex Sabena S-58s added to the Koksijde SAR flight after withdrawal from commercial service. B-12 was replaced by a Sea King and placed in storage pending disposal late in 1976. *(W.J. Bushell)*

Sikorsky S-58 B-7 OT-ZKG was one of five Sud-built HSS-1 helicopters supplied for SAR use. The anchor over the roundel shows the aircraft to be operated by the Zeemacht (Naval) Flight, but it was ditched in January 1976 and written off. *(via A.D. Annis)*

Stampe SV-4c V-64 The EPE/EVS aerobatic team of two Stampes was well known for its displays of 'mirror flying' during the late sixties. Appropriately painted black and white, V-64 had the front cockpit faired over when photographed in August 1966. *(MAP)*

Stampe SV-4b V-4 displays c/n 1144 on the fin, apparently out of sequence, as this should belong to V-2. Photographed at Brustem in 1971. *(via A. D. Annis)*

Sud Alouette II A2
The first Alouette II helicopters to be delivered to the Landmacht were A1 and A2, though both were written off in September 1960, before they could adopt the new style serial presentation. *(MAP)*

Sud Alouette II OL-A17 wearing its serial on the cockpit side in the style used during the sixties and early seventies, although the career of this particular example ended abruptly in August 1976. *(MAP)*

Sud Alouette II A30 demonstrates the current reversion to 'A' type serials only, having spent most of its career as OL-A30. Forty-two examples of the Artouste engined variant were delivered to Landmacht squadrons in Europe and Africa. *(J.-L. Gaynecoetche)*

Sud Alouette Astazou OL-A79 of the 'Blue Bees' stunt team seen at Middle Wallop in July 1971 is now serialled A79. Forty-two Alouette Astazou augmented the earlier Artouste-engined helicopters. *(W.J. Bushell)*

Sud Alouette IIIs
These two photographs illustrate clearly the dual-language situation obtaining in Belgium, with the helicopters marked *'Force Navale Belge'* to port, and *'Belgische Zeemacht'* to starboard. M-1/OT-ZPA - *(MAP)* M-2/OT-ZPB - *(via A.D. Annis)*

Swearingen Merlin 3A CF-02 photographed at Kleine Brogel in June 1976, derives its serial prefix 'F' from the fact that Swearingen's parent company is Fairchild. Six Merlins were delivered to 21 Smaldeel as Pembroke replacements in 1976. *(S.G. Richards)*

Piper L-18C Super Cub LX-FAA is one of three used by the Luxembourg Army since the early fifties for air observation and liaison. *(MAP)*

Supermarine Spitfire F.XIV SG-27 MN-P became a monument at St. Trond after suffering damage in an accident in July 1952. A total of 132 Spitfire XIV was delivered between 1947 and 1951. *(MAP)*

Supermarine Spitfire F.IX SM-15 c/n CBAF IX 1301 is ex RAF MJ783, and was delivered to Belgium in February 1948. It is now preserved in the Musée de l'Armée in Brussels, and has been repainted with code GE-B. *(Tony Lloyd)*

Westland Sea King Mk.48 RS-02 seen on a test flight as G-17-2 early in 1976. Five helicopters of this type now serve with 40 Smaldeel. *(Westland Aircraft)*

General Dynamics F-16 01567 seen at Alconbury in June 1975 gives a taste of the future Belgian equipment. The type will begin to replace the Starfighter from 1979. *(S.G. Richards)*

APPENDIX D

Preserved Aircraft

A second chance to see aircraft no longer in service is afforded by the Luchtmacht in preserving several gate guardians etc. The following may be seen –

Base	Aircraft type	Serial
Bevekom	Spitfire Starfighter	SG-31 'FR-34'
Bierset	Thunderflash	FR-34
Brustem	Meteor Stampe SV-4C Hunter	EG-79 V-46 'IF-70'
Chièvres	Meteor	EG-18
Dinant	Meteor	EG-162
Florennes	Spitfire	SG-57
Kleine Brogel	Thunderjet	FS-2
Koksijde	Hunter Dakota	ID-123 K-31
Maastridit	Thunderstreak	FU-108
Saffraanberg	Spitfire	SM-29
Zellick	Hunter Stampe SV-4C	ID-26 V-57

By far the greatest collection of aircraft is that in the Musée de l'Armée at Brussels, for which advance permission is required to view the sixty-plus exhibits in store/restoration here. Items of note include –

Aircraft type	Serial(s)/Remarks etc
Stampe SV-4B	OO-ATD, ex-RAF MX457, also V-28, V-56 and V-64
Spitfire	SG-37 and SM-15
Harvard	H-21 and H-39
Mosquito	MB-24
Hurricane	'LF345'
Proctor	P-4
Oxford	O-16
Super Cub	OL-L87
Hunter	ID-44 and ID-46
Thunderjet	FZ-132
Meteor	EG-224 and EG-247
Auster AOP.6	A-3, A-7, A-8, A-9, A-11, A-15, A-16, A-17 and A-22
Canuck CF-100	18534 (ex-RCAF)
Grunau Baby	PL-37
Schulgleiter SG38	PL-21
Thunderstreak	FU-30
Thunderflash	FR-28
Dakota	K-16
C-119G	CP-14
Sikorsky S-58	B-13

APPENDIX E

Belgian Military Radio Callsigns

For many years Air Force transports used five character radio callsigns in the OTxxx range with Army aircraft using OLxxx etc. Usually these were prominently displayed on the aircraft; far more so than the serial in most cases! Details of these will be found in the individual aircraft notes.

The first Air Force transports not to be allotted this type of callsign were the C-130H Hercules; these employed the callsigns 'Belgian Air Force 01' to 12'. In October/ November 1975, other types adopted callsigns within the same system; in each case these were allotted in numerical order and usually related directly to the aircraft's serial. Known ranges are –

'BAF'	21 and 22	Boeing 727s	CB-01 and CB-02
	31 and 32	Falcon 20s	CM-01 and CM-02
	41 to 43	HS.748s	CS-01 to CS-03
	51 to 56	Merlins	CF-01 to CF-06
	61 to 64	DC-6s	KY-1 to KY-4
	71 and 72	C-47s	K-8 and K-10
	81 to 92	Pembrokes	RM-01 to RM-12

From 1974, the Army adopted callsigns in the OTAxx ranges as follows –

OTAAA etc	Alouette II
OTACA etc	Alouette Astazou
OTALA etc	FBN Islander
OTAMA etc	Dornier 27

Like the Air Force ranges, callsigns are allotted in serial order, but omitting aircraft no longer in service; further details are included in the individual aircraft notes. Often the aircraft are coded with the 'last two' of this callsign.

The Rijkswacht helicopters follow the Army system, in the following OTGxx ranges –

OTGCA etc	Alouette Astazou
OTGIA etc	Aerospatiale Puma

APPENDIX F

Gliding Clubs

A fleet of gliders is maintained for the recreational use of Air Force personnel and Air Cadets at all bases in Belgium.

Known types are —

PL-1 to PL-7	Rhonlerche
PL-10 to PL-13	Schleicher Ka-2
PL-21	Schulgleiter SG-38 (wfu)
PL-30 to PL-33	Grunau Baby
PL-51	Schleicher Ka-6CR
PL-52 to PL-53	Schleicher Ka-6/8
PL-54	Schleicher Ka-8
PL-55 to PL-58	Schleicher Ka-8B
PL-59	Schleicher Ka-6E
PL-60	not known
PL-61 to PL-65	Schleicher ASK-13
PL-66 to PL-69	not known
PL-70	Fournier RF-4

Airspeed Horsas (presumably ex-RAF) were used for para-troop training at Schaffen; serials included —

PL-1808
PL-1809
PL-1810
PL-1812
PL-1813
PL-1815

APPENDIX G

Luxembourg

One of Europe's smaller nations, the Grand Duchy of Luxembourg gained independence from the German Confederation in 1867, and although economically united with Belgium since 1921, has continued to provide for its own defence.

Luxembourg has an area of 999 square miles and a population of 360,000. There is no air force (or navy!) but the army consists of 550 regular soldiers - all volunteers. Principal equipment is the 106mm recoilless rifle, 81mm mortar and TOW anti-aircraft guided missile.

Three Piper Super Cubs were attached to the army in 1952-3 for air observation and liaison —

Model	C/n	Ex-USAF	Registration
L-18C	18-3236	53-4836	LX-FAA
PA-18-95	18-2138	52-2538	LX-FAB
PA-18-95	18-2132	52-2532	LX-FAC

Of the above, LX-FAA crashed whilst exercising over the Luxembourg military practise area (in Belgium), and the other two were sold to 'Aero-Sport' - a civilian firm in Luxembourg - in 1969, and subsequently went to Lichenstein for the use of the local aero club. No replacements have been obtained.

APPENDIX H

Keeping up-to-date

Enthusiasts of squadron and serial-number data will find much to interest them in the monthly publications listed hereunder. These cover developments throughout Europe, but regularly contain information from Benelux which will be of assistance in keeping this book up-to-date.

British Aviation Review
8 Nightingale Road
Woodley
Berkshire
England

'Flash' Aviation Magazine
P.O. Box 855
Eindhoven
Holland

South East Air Review
18 Green Lawns
Southbourne Gardens
Eastcote, Ruislip
Middlesex
England

Any reader not presently subscribing to these publications is advised to put down this book *now* and write to the addresses given for further details.

Many other excellent publications have not been included in the above by reason of the fact that they are not *regular* publishers of Benelux serial-number information.

Aviation Books

We have pleasure in presenting below a summary of our current stock list of aviation titles. Whilst this is correct at the time of publication, it is inevitable that there will be additions, deletions, changes of price and specification from time to time. We therefore reserve the right to alter the terms of our offers as circumstances dictate. A comprehensive book list is available, which includes more details of the books shown below, and there are also issued 'Booknews' leaflets with details of new books etc. Please contact us for further details.

Patrick Stephens/Kookaburra

Bombing Colours 1914-37 £ 3.50
Bombing Colours 1937-73 £ 4.50
Fighting Colours 1937-75 £ 3.95
Airfix Magazine Guides -
11 RAF Camo. of WWII £ 1.40
14 American Fighters WWII £ 1.40
18 USAAF Camo. of WWII £ 1.40
Night Fighters - Gunston £ 4.50
Luftwaffe Camouflage & Markings 1939-45 - Smith & Gallaspy - in three volumes each at £ 8.50
Aces & Pilots of the 9th, 12th & 15th USAAF £ 7.50
RNZAF in Pacific 1939-45 £ 4.50

Faber & Faber

2 Group RAF 1936-45 £ 8.75
Mosquito - Bowyer £ 7.00

Harleyfords

Air Aces 1914-18 War £ 7.50
Aircraft Markings of the World 7.50
de Havilland Tiger Moth £ 3.75
Fighter Aircraft 1914-18 £ 9.50
Focke-Wulf FW.190 £ 7.50
Avro Lancaster £ 7.50
Marine Aircraft 14-18 War £ 7.50
Messerschmitt Bf.109 £ 7.50
Russian Civil & Military A/c £ 9.00
Spitfire £ 7.50
Von Richthofen £ 9.50
US Navy & MC Fighters 18-62 9.50
US Army & AF Ftrs.1916-61 £ 9.50
Aircraft Camo.& Mkgs '07-54 £ 9.50

Ian Allan

Hurricane Special - Allward £ 2.75
Hampden Special - Bowyer £ 2.50
Spitfire Special - Hooton £ 1.95
Wellington Special - Lumsden £ 1.95
Beaufort Special - Robertson £ 2.50
B-24 Liberator - Allen Blue £ 4.95
Attack Aircraft of the West £ 3.95
Military A/c of the World £ 2.95
Early Supersonic Fighters of the West - Gunston £ 4.95
Photo Reconnaissance £ 5.25
Air War over France 1939-40 £ 3.30
Air War over Korea £ 2.95
Air War over Spain £ 6.50
Helicopters of the World £ 2.95
Strategic Air Command £ 4.95
Beaufighter at War £ 4.95
Mosquito at War - Bowyer £ 3.50
Hurricane at War - Bowyer £ 4.50
Sunderland at War - Bowyer £ 4.95
Mustang at War - R.Freeman £ 4.75
Thunderbolt at War - W.Hess £ 5.50
Spitfire at War - A.Price £ 4.50
Lancaster at War £ 3.95
Battle over the Reich £ 3.95
RAF Bomber Command and its aircraft 1936-40 -Moyes £ 5.95
RAF Jet Bomber Flypast £ 1.65
RAF Jet Fighter Flypast £ 1.80
Typhoon & Tempest at War £ 4.95
Aircraft of World War II £ 2.50
Fortress at War - R.Freeman £ 5.50
Blitz over Britain - A.Price £ 4.75
Air Defence of Gt.Britain £ 3.30
Pathfinders at War.- Bowyer £ 5.50
Swordfish Special - Harrison £ 2.75

Macdonald & Janes

Bomber in WW2 - A.Price £ 4.95
Attack Warning Red - Wood £ 8.50
Airships - Henry Beaubois £ 15.00
World Combat Aircraft Directory - Norman Polmar £ 6.95
Early Aviation at Farnborough -
Vol.1 Balloons /Airships £ 6.50
Vol.2 First Aeroplanes £ 8.95
Instruments of Darkness £ 5.95
Wings of the Luftwaffe £ 5.95
Soviet Air Force since 1918 £ 6.50
Pathfinder Force (8 Group) £ 6.95
Spitfire - A.Price (due 9/77) £ 5.95
Project Cancelled - D.Wood £ 6.50
World Defence Who's Who £ 8.50
Warplanes of Third Reich £ 12.50
Mighty Eighth - R.Freeman £ 6.50
U S Strategic Bomber £ 3.25
Ground Attack Aircraft £ 4.95
Bomber Sqdns.of the RAF £ 12.50
Fighter Sqdns.of the RAF £ 12.50
Warplanes of First World War series-
1 British Fighters A-P £ 1.25
3 British Fighters T-Z £ 1.25
4 French Fighters A-G £ 1.25
5 French Fighters H-M £ 1.25
Warplanes of 2nd World War series -
1 Fighters Australia-Germany 1.15
3 Fighters Japan-Russia £ 1.15
4 Fighters USA-Yugoslavia £ 1.15
5 Flying Boats £ 1.15
7 Bombers Australia-Germany 1.15
9 Bombers & Recce.Germany 1.15
Janes Pocket Books (PVC covers) -
2 Major Combat Aircraft £ 2.25
6 Light Aircraft £ 2.25
7 Airship Development £ 2.75
12 Experimental Aircraft £ 2.25
14 Homebuilt Aircraft £ 2.75
10 Missiles £ 2.25
13 RPVs/Robot Aircraft £ 2.75
15 Record Breaking Aircraft £ 2.75

Terence Dalton

Martlesham Heath- Kinsey £ 4.80
Aviation - Flight over the Eastern Counties since 1937 £ 5.80
Both Feet in the Air -Jackson £ 4.80

Janes Yearbooks

Janes All the Worlds Aircraft 1976/7 edition £ 25.00
Fighting Ships, Weapon Systems, Railways etc. all available to special order - details on request.

Aviation Archaeology

Aviation Archaeology (PSL) - Bruce Robertson £ 4.50
Warplane Wreck Inv.Group books :
High Ground Wrecks £ 1.00
Mosquito Crash Log £ 1.25
Beaufighter Crash Log £ 1.25
Sunderland Crash Log £ 1.75

Guinness

History of Air Warfare £ 6.50
Book of Air Facts & Feats £ 6.50

Aero Publishers Inc.

L-1011 TriStar & Lockheed £ 8.00
747 - Boeing Super Jet £ 8.00
US Fighters 1925-1980s £ 9.00
US Bombers 1928-1980s £ 7.50
Italian Civ.&Mil.A/c 1930-45 £ 7.50
Ninth AF in WW2 - K.Rust £ 9.50
Tin Goose (Ford Trimotor) £ 2.50
US Civil Aircraft - Joseph Juptner - in six volumes each at..... £ 7.50
Grumman F-14 Tomcat £ 4.50
General Dynamics F-16 £ 4.50
Convair F-106 Delta Dart £ 4.50
Republic F-105 Thunderchief £ 4.00
Aero Pictorials -
1 Luftwaffe in WWII £ 2.50
2 USAAF in the Pacific £ 2.50
3 R.Australian & RNZAF £ 2.50
4 Flying Leathernecks WWII £ 2.50
World War II series -
Skybird Group (353rd FtrGp) £ 2.50
Fortunes of War (492 Bomb) £ 2.50
Fireball Outfit (457 BombGp) £2.50
Checkertail Clan (325 FtrGp) £ 2.50
Yoxford Boys (357 Ftr.Gp) £ 2.50

US Naval Institute Press

Dauntless Dive Bomber £ 10.95

Squadron/Signal

1 Luftwaffe part 1 £ 2.25
2 Luftwaffe part 2 £ 2.25
3 Luftwaffe Bombers 1 £ 2.25
4 Luftwaffe Bombers 2 £ 2.25
5 F-4 Phantom £ 2.25
6 Heinkel 111 £ 2.25
7 F-8 Crusader £ 2.25
8 Luftwaffe part 4 £ 2.25
9 F-100 Super Sabre £ 2.25
10 Junkers 52 £ 2.25
11 A-4 Skyhawk £ 2.25
12 Flying Fortress £ 2.25
13 The New Luftwaffe £ 2.25
14 Gunslingers £ 2.25
15 F-106 Delta Dart £ 2.25
16 Junkers 88 £ 2.25
17 F-105 Thunderchief £ 2.25
18 P-47 Thunderbolt £ 2.25
19 Focke-Wulf 190 £ 2.25
20 A-6 Intruder £ 2.25
21 B-24 Liberator £ 2.25
22 A-7 Corsair £ 2.25
23 B-52 Stratofort ress £ 2.25
24 F-15 Eagle £ 2.25
25 P-38 Lightning £ 2.25
26 Curtiss P-40 £ 2.25
27 F-104 Starfighter £ 2.25
28 B-47 Stratojet £ 2.25
30 Messerschmitt Bf.110 £ 2.25
And Kill Migs - Drendel £ 3.75
Regia Aeronautica 1940-43 £ 2.95
Armée de l'Air 1937-45 £ 2.95
Phantom II - Lou Drendel £ 3.25

Hamlyn

Encyc.Worlds Combat Aircr. £ 4.95
Modern Combat Aircraft £ 0.75
Helicopters at War £ 2.95
Naval Aircraft £ 2.95
World of Aviation £ 3.95
Encyc.of Air Warfare £ 4.95
Soviet War Machine £ 4.95
Seaplanes & Flying Boats £ 3.50
History of the RAF £ 4.95
USAF in WW2 (due 11/77) £ 4.95
Air Forces - WW1 & WW2 (do) 5.95
Fighters 1914-45 - Gunston £ 2.95

OUR OWN PUBLICATIONS

GERMAN MILITARY AVIATION 1956-1976 - Paul A. Jackson
ompanion volume to this book, with similar style and content.
dback edition £ 4.95 Softback edition £ 2.95

ANISH AND PORTUGUESE MILITARY AVIATION - John Andrade
so in the same series. For publication October 1977.
ardback edition £ 4.95 Softback edition £ 2.95

BEAGLE AIRCRAFT - A PRODUCTION HISTORY £ 0.90

LOCKHEED AEROBATIC TROPHY 1955-65 - Tony Lloyd £ 1.00

All the prices above are post-free. Please note that 'French Military Aviation' is now out of print. A new edition is planned for mid-1978.

U.S. CIVIL AIRCRAFT NEWS Bi-monthly illustrated magazine with news, manufacturers notes etc., plus a sixteen page feature devoted to extracts from the current US register, covering more interesting types - Either 50p per issue, or £ 3 for annual subscription.

Frederick Warne

Observers Bk.Aircraft 1977 £ 1.10
Basic Civil A/c Directory £ 2.95
Basic Military A/c Directory £ 2.95
Soviet Aircraft Directory £ 3.50
World Airlines/Airliners Dir. £ 4.95

Putnam

Aeroflot:Soviet Air Transport £ 6.50
Russian Aircraft since 1940 £10.00
Soviet Transport A/c s.1940 £ 3.15
British Fighter since 1912 £ 4.50
British Bomber since 1914 £ 4.50
British Avn:Gt.War/Armistice £ 6.50
British Avn:Adventuring Yrs £ 10.00
German Giants - R-Planes £ 6.50
German Aircraft of WWI £ 5.25
German Aircraft of WW2 £ 9.50
The Captive Luftwaffe £ 6.00
US Military A/c since 1908 £ 7.50
US Navy A/c since 1911 £ 10.50
A/c of the RAF since 1918 £ 8.50
Polish Aircraft 1893-1939 £ 8.50
Armament of British A/c £ 5.50
Airlines of U.S since 1914 £ 7.50
Airspeed A/c since 1931 £ 3.25
Blackburn A/c since 1909 £ 6.50
Armstrong Whitworth s.1913 £ 6.00
Bristol A/c since 1910 £ 8.50
Fairey A/c since 1915 £ 8.50
Gloster A/c since 1917 £ 8.50
Handley Page Aircraft £ 12.50
Hawker A/c since 1920 £ 5.00
Vickers A/c since 1908 £ 6.50
Miles Aircraft since 1925 £ 8.50
de Havilland Aircraft since 1909 (2nd edn. due March 1978) £ 12.50
British Civil Aircraft since 1919 - A. J.Jackson. In three volumes :
Vol.1 ABC- Chilton £ 8.50
Vol.2 Chrislea to HS.650 £ 8.50
Vol.3 HS.748 to Zlin £ 8.50

Aircam

1 NA P-51D Mustang £ 1.95
2 Rep.P-47 Thunderbolt £ 1.95
4 Spitfire Mks. I - XVI £ 1.95
5 NA P-51B/C Mustang £ 1.95
6 Curtiss Kittyhawk I - IV £ 1.95
8 Spitfire Mk. XII-24 £ 1.95
10 Lockheed P-38 Lightning £ 1.95
12 Avro Lancaster £ 1.95
14 F/RF-84F Thunderstreak £ 1.95
15 B-17B-H Flying Fortress £ 1.95
17 NA F-86A-L Sabre £ 1.95
23 Vought F4U-17 Corsair £ 1.95
24 Hawker Hurricane I/IV £ 1.95
26 Hawker Hunter F1/T66 £ 1.95
27 Douglas A-4 Skyhawk £ 1.95
28 de Havilland Mosquito £ 1.95
30 F-4 Phanton II Vol.1 £ 1.95
31 Vought F-8 Crusader £ 1.95
33 de Havilland Vampire £ 1.95
37 EE/BAC Lightning £ 1.95
39 Messerschmitt 109 Vol.1 £ 1.95
40 Messerschmitt 109 Vol.2 £ 1.95
41 F-4 PhantomII Vol.2 £ 1.95
42 Messerschmitt 109 Vol.3 £ 1.95
43 Messerschmitt 109 Vol.4 £ 1.95
44 Focke-Wulf 190A/F/G £ 1.95

Specials -
S1 Battle of Britain £ 1.95
S2 Finnish AF 1918-68 £ 1.95
S3 Sharkmouth 1916-45 (1) £ 1.95
S4 Sharkmouth 1945-70 (2) £ 1.95
S5 Czech Air Force 1918-70 £ 1.95
S6 Luftwaffe Cols/Mkgs 39/45 1.95
S8 -ditto - Vol.2 £ 1.95
S19 -ditto- Vol.3 £ 1.95
S10 Luftwaffe Bomber Camo. & Markings 1940 £ 1.95
S11 Luftwaffe Fighter-Bomber Cam/Marks 1940 £ 1.95
S13 USAAF Heavy Bomb Gp 1.95
S14 -ditto- Vol.2 £ 1.95
S17 50 Fighters 1938-45 (1) £ 1.95
S18 50 Fighters 1938-45 (2) £ 1.95
S12 Aerobatic Teams 1950-71 1.95

Aircam/Airwar series -
1 RAF Fighters, Europe 39-42 1.95
2 USAAF Heavy Bomber Units ETO & MTO 1942-45 £ 1.95
3 Spanish Civil War Forces £ 1.95
4 L'waffe Grd.Attack Units £ 1.95
5 RAF Bomb Units 39-42 £ 1.95
6 L'waffe Ftrs, Europe 39-41 £1.95
7 USAAF Medium Bomber Units ETO & MTO 1942-45 £ 1.95
8 USAAF Ftrs, Europe 42-45 1.95

Military Aviation Review

Monthly illustrated magazine for military aviation fans. Issues 1 to 3 (May-July 76) each 40p. Issues 4 onwards each at 50p.

Airline Publications

World Airline Fleets 1977 £ 3.30
World Airline Colours Vol.1 £ 3.30
Heathrow Schedules £ 0.66
Gatwick Schedules £ 0.42
Biz-Jet 77 - Brian Gates £ 0.95

Other Books

Fighter - the true story of the Battle of Britain - Len Deighton £ 4.95
Combat Aircraft of WW2 (Arms & Armour Press - Elke C.Weal £ 9.95
Airborne - Neil Williams £ 4.95
Fields of Little America (illus.hist.of 8th AF, 2Div 1942-45) Due 11/77 - Softback £ 3.50 or hardback £ 5.25
Sweden - Haven of Refuge £ 3.20

Enthusiasts publications

12 Years On (Islander/Trisl.) £ 1.80
Lockheed Hercules 1955-77 £ 1.80
dH Venom - Roger Lindsay £ 0.95
Gloster Javelin 1-6 - Lindsay £ 1.60
US Naval Aviation Today £ 2.75
JP Airline Fleets 1977 £ 3.30
Dutch Civil Register 1977 £ 1.30
CSAG Scotland Scanned 77 £ 1.20
HP.137 Jetstream £ 0.65
Warpaint 1 - F/RF-84F £ 0.95
Warpaint 2 - EE Lightning £ 0.95
Merseyside publications -
Wrecks & Relics (5th edn) £ 1.50
British Isles Airfield Guide £ 0.50
British Gliders (2nd edn) £ 0.65
US Navy Serials 1941-76 £ 1.30
USAF Serials 1946-77 £ 1.60
British Museum Aircraft £ 2.00
British Air Arms ?
British Independent Airlines -
1 ACE to British Westpoint £ 2.00
2 Brooklands to Irelfly £ 2.00
3 Irish Air Ch. to Sagittair £ 2.00
4 Scillonia to Yorkshire AC £ 2.00
OR a full set post included at £ 9.00
Air-Britain registers -
1977 UK & Eire £ 2.25
1977 France £ 2.70
1977 Australia, NZ, Pacific £ 2.70
1977 Germany ?
1976 Scandinavia £ 1.35
1976 Africa (57 countries) £ 2.10
1976 Southern Europe/M.East 1.80
1975 Benelux,Swiss,Austria £ 1.35
World Airline Fleets H/b 77 £ 2.25
RAF Serials K1000-K9999 £ 3.00
Dove/Heron monograph £ 1.10
Viking/Varsity/Valetta £ 1.35
Boeing 727 £ 1.60
Britannia/B.170/CL-28/CL-44 £ 2.40

POST AND PACKING Please add 10% to all orders to cover second class postage or parcel post costs in the UK and Eire, or surface mail at 'bookpost' rates elsewhere. If airmail is required, please add 50%. ORDERS OVER £ 15 will, however, be sent post free.

PAYMENT is requested with order, please, by cheque, postal order or IMO payable to 'Midland Counties Publications'.

OVERSEAS CUSTOMERS please pay by IMO, or sterling cheque drawn on a UK bank. At the senders risk we also accept currency notes. US and Canadian customers may pay by personal dollar check, but add 10% please to cover our bank negotiation charges. At the time of going to press, the dollar exchange rate is 1.74, so that a convenient way of converting, to include postage and bank charge also, is merely to double the sterling book price, and call it dollars.

MIDLAND COUNTIES PUBLICATIONS
17 Woodstock Close, Burbage, Hinckley, Leicestershire LE10 2EG